Sustainable Construction II

Sustainable Construction II

Editors

Víctor Yepes
José V. Martí

MDPI • Basel • Beijing • Wuhan • Barcelona • Belgrade • Manchester • Tokyo • Cluj • Tianjin

Editors
Víctor Yepes
Institute of Concrete Science and Technology (ICITECH),
Universitat Politècnica de València
Spain

José V. Martí
Institute of Concrete Science and Technology (ICITECH),
Universitat Politècnica de València
Spain

Editorial Office
MDPI
St. Alban-Anlage 66
4052 Basel, Switzerland

This is a reprint of articles from the Special Issue published online in the open access journal *Sustainability* (ISSN 2071-1050) (available at: https://www.mdpi.com/journal/sustainability/special_issues/Sustainable-Construction).

For citation purposes, cite each article independently as indicated on the article page online and as indicated below:

LastName, A.A.; LastName, B.B.; LastName, C.C. Article Title. *Journal Name* **Year**, *Volume Number*, Page Range.

ISBN 978-3-0365-0484-1 (Hbk)
ISBN 978-3-0365-0485-8 (PDF)

Contents

About the Editors

Víctor Yepes is a full professor of Construction Engineering; he holds a Ph.D. in civil engineering. He serves at the Department of Construction Engineering, Universitat Politecnica de Valencia, Valencia, Spain. He has been the Academic Director of the M.S. studies in concrete materials and structures since 2007 and a Member of the Concrete Science and Technology Institute (ICITECH). He is currently involved in several projects related to the optimization and life-cycle assessment of concrete structures, as well as optimization models for infrastructure asset management. He currently teaches courses in construction methods, innovation, and quality management. He has authored more than 250 journals and conference papers, including more than 100 published in journals quoted in JCR. He acted as an expert for project proposal evaluation for the Spanish Ministry of Technology and Science, and he is a main researcher in many projects. He currently serves as an Editor-in-Chief for the *International Journal of Construction Engineering and Management* and a member of the editorial board of 12 other international journals (*Structure and Infrastructure Engineering, Structural Engineering and Mechanics, Mathematics, Sustainability, Revista de la Construcción, Advances in Civil Engineering, Advances in Concrete Construction*, among others).

José V. Martí is an Associate Professor in the Department of Construction Engineering and Civil Engineering Projects at the Universitat Politècnica de València, Spain. Initially, he worked for private companies in the construction sector, business consulting, and financial entities, and later as a freelance professional. He has taught since 1995 and, in many cases, served as the head of subjects in the Master of Civil Engineering, Geodetic Engineering and Topography, and in the degrees in Civil Engineering and Public Works. He has educated students on all matters related to the subject of construction procedures, quality, organization of works, and civil engineering machinery. He has participated in nine didactic books, 23 notebooks, 31 articles in teaching congresses, and a teaching innovation project. For his own research activity, he has a book as an author, a book chapter, and a participant in 29 articles in JCR journals. His lines of research are mainly focused on the optimization of structures through the application of metaheuristic techniques, and on the life-cycles and sustainability of structures.

Preface to "Sustainable Construction II"

Construction is one of the main sectors that generates greenhouse gases. This industry consumes large amounts of raw materials, such as stone, timber, water, etc. Additionally, infrastructure should provide service over many years without safety problems. Therefore, their correct design, construction, maintenance, and dismantling are essential to reducing economic, environmental, and societal consequences. That is why promoting sustainable construction has recently become extremely important.

To help address and resolve these types of questions, this book is comprised of five chapters that explore new ways of reducing the environmental impacts caused by the construction sector, as well to promote social progress and economic growth. The chapters collect papers included in the "Sustainable Construction II" Special Issue of the *Sustainability* journal.

We would like to thank both the MDPI publishing and editorial staff for their excellent work, as well as the 18 authors who collaborated in its preparation. The papers cover a wide spectrum of issues related to the use of sustainable materials in construction, the optimization of designs based on sustainable indicators, the life-cycle assessment, the decision-making processes that integrate economic, social, and environmental aspects, and the promotion of durable materials that reduce future maintenance.

Víctor Yepes, José V. Martí
Editors

sustainability

MDPI

Article

Environmental and Social Impact Assessment of Optimized Post-Tensioned Concrete Road Bridges

Vicent Penadés-Plà [1], David Martínez-Muñoz [1], Tatiana García-Segura [2], Ignacio J. Navarro [2] and Víctor Yepes [1,*]

[1] Institute of Concrete Science and Technology (ICITECH), Universitat Politècnica de València, 46022 Valencia, Spain; vipepl2@cam.upv.es (V.P.-P.); damarmu1@cam.upv.es (D.M.-M.)

[2] Department of Construction Engineering and Civil Engineering Projects, Universitat Politècnica de València, 46022 Valencia, Spain; tagarse@upv.es (T.G.-S.); ignamar1@cam.upv.es (I.J.N.)

* Correspondence: vyepesp@cst.upv.es; Tel.: +34-9638-79563; Fax: +34-9638-77569

Received: 3 April 2020; Accepted: 20 May 2020; Published: 22 May 2020

Abstract: Most of the definitions of sustainability include three basic pillars: economic, environmental, and social. The economic pillar has always been evaluated but not necessarily in the sense of economic sustainability. On the other hand, the environmental pillar is increasingly being considered, while the social pillar is weakly developed. Focusing on the environmental and social pillars, the use of methodologies to allow a wide assessment of these pillars and the integration of the assessment in a few understandable indicators is crucial. This article is structured into two parts. In the first part, a review of life cycle impact assessment methods, which allow a comprehensive assessment of the environmental and social pillars, is carried out. In the second part, a complete environmental and social sustainability assessment is made using the ecoinvent database and ReCiPe method, for the environmental pillar, and SOCA database and simple Social Impact Weighting method, for the social pillar. This methodology was used to compare three optimized bridges: two box-section post-tensioned concrete road bridges with a variety of initial and maintenance characteristics, and a pre-stressed concrete precast bridge. The results show that there is a high interrelation between the environmental and social impact for each life cycle stage.

Keywords: sustainability; LCA; S-LCA; social assessment; ecoinvent; SOCA

1. Introduction

Since its definition in 1987 by the World Commission on Environment and Development [1], the concept of sustainability has been attracting increasing attention in many sectors of our society. However, it was not until 2015 that the first set of Sustainable Development Goals was established as a response to growing social needs and environmental degradation [2].

Despite social assessment being an important part of the sustainability definition, its evaluation is underestimated or relatively weak with respect to the other pillars of sustainability when sustainability assessments of products, processes, or services have been carried out [3,4]. Vallance et al. [5] stated that this is due to the fact that the definition of social sustainability is quite ambiguous, and Murphy [3] indicated that there are no clear criteria for assessing sustainability. However, social equity, education, basic health, and participatory democracy are important for sustainability development [6]. At present, there is a trend toward giving the social pillar the same importance as the economic and environmental pillars [7–10]. This is demonstrated by the fact that 6 of the 17 sustainable development goals proposed by the United Nations focus on social problems.

The complex stakeholder situation in construction projects makes performing a social sustainability assessment difficult [11]. Valdes-Vasquez and Klotz [12] indicated that projects in the construction sector involve clients, employees, the community, and industry, and have the intention of satisfying current and future needs. Later, Almahmoud and Doloi [13] stated that the social aspect in the construction sector can be represented through the satisfaction of the different stakeholders involved in the projects, including industry, users, and the community. They also indicated that the importance of the impact of the project for future generations and the impact on present generations through health, safety, and conditions of workers must be taken into account.

A sustainability assessment becomes a decision-making problem [14–17]. In addition, this decision-making problem can be assessed taking into account the different perspectives of the decision-makers reaching robust sustainable solutions [18]. Bridges have been widely investigated from the technical point of view [19–21]. This provides a great number of different designs of bridges that engineers must select from a sustainable point of view. Penadés-Plà et al. [22] reviewed the criteria considered to assess the different pillars of sustainability in bridges, as well as the multi-attribute decision-making methods used to obtain a global sustainability assessment. This review shows that the economic pillar is the most developed pillar. Although some early works only studied the initial cost of the bridge, a life cycle cost assessment (LCCA) is, nowadays, widely used. Conversely, a life cycle assessment (LCA) is less common. For the environmental part, few studies have applied environmental life cycle assessments (E-LCAs) to bridges. Horvath and Hendrickson [23] and Widman [24] conducted the first investigations. Then other works followed. However, most of them did not consider all phases of the bridge's life cycle [25,26], or they focused on a limited number of environmental indicators (usually energy and CO_2) [27,28]. It was not before the study by Steele et al. [29] that a full E-LCA was performed. Pang et al. [30] compared different bridge maintenance operations, and Du et al. [31] and Hammervold et al. [32] compared several bridge designs. Regarding the social part, there is no consensus to define the criteria that best represent social life cycle assessment (S-LCA). Some works have considered criteria as divergent as detour time, dust, or noise [33–35].

In this paper, a bibliographic review of the LCA methods, both environmental and social, will first be conducted in Section 2. After that, Section 3 explains the methodology used, after discussing the best methods to assess the social and environmental pillars of bridges. In Section 4, these methods are used to carry out a sustainability assessment of three road bridges: two box-section road bridges with different initial and maintenance characteristics, and a pre-stressed concrete precast bridge. Section 5 shows the results of all the pillars of sustainability, focusing on social assessment. As a final point, conclusions are presented in Section 6.

2. Life Cycle Assessment Methods

To carry out a complete sustainability assessment, it is essential to consider the whole life cycle of a product, service, or process. This is even more important in the construction sector, because structures are built to provide a service over a long time, and therefore the assessment of the use and maintenance stage becomes quite important. For this purpose, a life cycle assessment methodology is used. At this point, it is necessary to point out that despite the LCA techniques—used to assess both the environmental and social pillars—having the same central core, there are some differences between them. For this reason, in this study, the term LCA will be used when referring to the common trunk of the technique and the terms E-LCA (environmental pillar) and S-LCA (social pillar) are going to be used for specific assessments.

Focusing on environmental, the ISO 14040 [36] defines LCA as a technique for evaluating the environmental aspect and impacts caused by a process, product, or service through a system of input flows (data) that cause output flows (impacts). The most common guide to carry out the social assessment [37] follows the same steps as this code. ISO 14040 [36] divides the LCA into four phases:

- Goal and scope definition
- Inventory analysis
- Impact assessment
- Interpretation of results

The impact assessment step of the LCA is crucial since the information obtained from the life cycle inventory is transformed into a set of understandable indicators. Due to the complexity of this transformation, some methodologies have been developed to simplify this step, called life cycle impact assessment (LCIA) methods. In this sense, the assessment and comparison between different cases become easier.

2.1. *Environmental Life Cycle Impact Assessment*

In the E-LCA, there are two approaches to transform the life cycle inventory into understandable indicators: the "midpoint approach" and the "endpoint approach". The midpoint approach refers to environmental impact, while the endpoint approach refers to environmental damage. The midpoint approach provides more complete information, and the endpoint approach allows for more concise information (Figure 1).

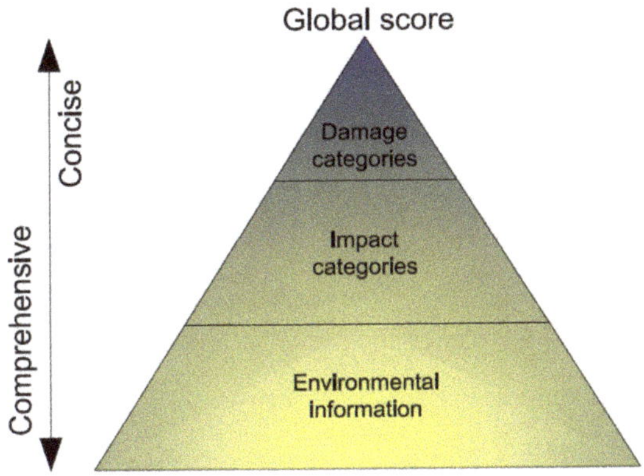

Figure 1. Environmental life cycle cost assessment (E-LCA) approaches.

Another way to understand the differences between these two approaches is to consider that the midpoint approach is the direct cause, while the endpoint approach is the long-term consequence. For example, any process, product, or service that affects climate change has gas emissions to the atmosphere that cause several environmental problems such as ozone depletion or global warming (midpoint approach); in the long-term approach, these gas emissions will cause damage to the ecosystem, human health, or resources. In this example, ozone depletion can lead to increased skin cancer problems (endpoint approach).

Table 1 shows the most common methods for each category and the indicators (midpoint indicators for the midpoint approach and endpoint indicators for the endpoint approach) considered for each E-LCIA method. Each approach uses different methods to convert environmental information into comprehensible and understandable indicators. Within midpoint approach methods, the classical methods are the CML [38], EDIP 2003 [39], and TRACI [40]. These methods provide a set of midpoint indicators that indicate the direct cause by a process, product, or service. The total number of these indicators is usually quite high, providing accurate information, but which is sometimes difficult to interpret. In addition, midpoint indicators are more difficult to understand because they depict an earlier stage in the cause–effect chain. Endpoint approach methods are damage-oriented methods, such as the Eco-indicator 99 [41], EPS [42], and eco-scarcity [43]. These methods provide a set of endpoint indicators that indicate the long-term consequences for a process, product, or service. The number of these indicators is usually quite small because it is an aggregation of the midpoint indicators. Therefore, there is a loss of detail and the information is not as accurate as in the case of midpoint methods, but much easier to interpret. In addition, there is a set of new methods, which combines the methods of midpoint and endpoint approaches, such as the ReCiPe [44,45], LIME [46], and IMPACT 2008 [47].

Table 1. Environmental life cycle impact assessment (E-LCIA) indicators.

E-LCIA Group	E-LCIA Method	Midpoint Indicators	Endpoint Indicators
Midpoint approach	CML 2000	Obligatory impact categories: Depletion of abiotic resources, climate change, land competition, stratospheric ozone depletion, human toxicity, freshwater aquatic ecotoxicity, terrestrial ecotoxicity, marine aquatic ecotoxicity, photo-oxidant formation, acidification, and eutrophication. Optional impact categories: Loss of life support function, loss of biodiversity, marine sediment ecotoxicity, freshwater sediment ecotoxicity, impacts of ionizing radiation, waste heat, malodorous air, noise, casualties, lethal, non-lethal, depletion of biotic resources, desiccation, and malodorous water	
	EDIP 2003	Global warming, ozone depletion, terrestrial eutrophication, acidification, aquatic eutrophication, photochemical ozone formation, ecotoxicity, human toxicity, and noise	
	TRACI	Ozone depletion, global warming, smog formation, eutrophication, acidification, eco-toxicity, human health cancer, human health non-cancer, human health criteria pollutants, and fossil fuel depletion	
Endpoint approach	EI99		Climate change, ozone layer depletion, acidification, eutrophication, respiratory effects, carcinogenicity, ionizing radiation, ecotoxicity, land use, mineral resources, fossil resources
	EPS		Life expectancy, severe morbidity and suffering, morbidity, severe nuisance, wood production capacity, nuisance crop production capacity, fish and meat production capacity, base cation capacity, production capacity for water, share of species extinction, depletion of element reserves, depletion of fossil reserves (coal), depletion of fossil reserves (gas), depletion of fossil reserves (oil), and depletion of mineral reserves
	Eco-scarcity		Ozone depletion, photochemical oxidant formation, respiratory effects, air emissions, surface water emissions, radioactive emissions, cancer caused by radionuclides emitted to the sea, emissions to groundwater, emissions to soil, radioactive wastes, landfill municipal (reactive) wastes, hazardous wastes (stored underground), water consumption, gravel consumption, primary energy resources, endocrine disruptors, and biodiversity losses

Table 1. *Cont.*

E-LCIA Group	E-LCIA Method	Midpoint Indicators	Endpoint Indicators
Midpoint/Endpoint approach	ReCiPe	Climate change, ozone depletion, terrestrial acidification, freshwater eutrophication, marine eutrophication, human toxicity, photochemical oxidant formation, particulate matter formation, marine ecotoxicity, terrestrial ecotoxicity, freshwater ecotoxicity, ionizing radiation, agricultural land occupation, urban land occupation, natural land transformation, water depletion, mineral resource depletion, fossil fuel depletion	Damage to human health, damage to ecosystem diversity, and damage to resource availability
	LIME	Ozone layer depletion, global warming, acidification, photochemical oxidant formation, regional air pollution, human-toxic chemicals, ecotoxic chemicals, eutrophication, waste landfill, land use, resources, and consumption	Cataracts, skin cancer, other cancers, respiratory diseases, thermal stress, infectious diseases, agricultural production, hypoalimentation, disaster causality, forestry production, fishery production, loss of land-use, energy consumption, user cost, terrestrial ecosystem, aquatic ecosystem
	IMPACT 2000+	Human toxicity, respiratory effects, ionizing radiation, ozone depletion, photochemical oxidant formation, terrestrial ecotoxicity, aquatic ecotoxicity, aquatic eutrophication, terrestrial eutrophication and acidification, land occupation, global warming, non-renewable energy, and mineral extraction	Damage to human health, damage to ecosystem quality, damage to climate change, and damage to resources

2.2. Social Life Cycle Impact Assessment

The social pillar of sustainability is the least studied and probably the most diffuse and weakest pillar of sustainability. However, for a complete sustainability assessment, it is necessary to obtain a complete set of social indicators that can be used to carry out an accurate comparison and assessment of alternatives. Currently, a simple Social Impact Weighting Method is used to convert the life cycle inventory into understandable indicators. However, there are two important social databases: PSILCA (Product Social Impact Life Cycle Assessment) [48] and SHDB (Social Hotspots Database) [49]. Both S-LCIA databases are inspired by UNEP/SETAC guidance [37] and use the activity variable "worker hour" in order to quantify the social impacts. Table 2 shows existing and planned categories of both methods grouped by stakeholders for the PSILCA database [50] and for the SHDB database [51]. However, the number of categories is lower nowadays.

Table 2. Social life cycle impact assessment (S-LCIA) categories.

S-LCIA Database	CATEGORIES
Product Social Impact Life Cycle Assessment (PSILCA)	WORKERS: Child labor, fair salary, discrimination, forced labor, health and safety, social benefits and legal issues, working time, workers' rights. VALUE CHAIN ACTORS: Corruption, promoting social responsibility, fair competition, supplier relationships. SOCIETY: Contribution to economic development, health and safety, prevention and mitigation of conflicts. LOCAL COMMUNITY: Access to material resources, respect of indigenous rights, local employment, safe and healthy living conditions, migration. CONSUMERS: Health and safety, transparency, end of life responsibility.
Social Hotspot Database (SHDB)	LABOR RIGHTS AND DECENT WORK: Child labor, discrimination, excessive working time, freedom of association, forced labor, labor laws, migrant labor, poverty, social benefits, unemployment, wage assessment. HUMAN RIGHTS: Indigenous rights, human health issues, gender equity, high conflicts. HEALTH AND SAFETY: Injuries and fatalities, toxins and hazards. GOVERNANCE: Legal system, corruption. COMMUNITY: Drinking water, children out of school, hospital beds, sanitation, smallholder vs. commercial farms.

The PSILCA database was developed by GreenDelta and presented in 2013. This database provides information to carry out the assessment of the social pillar of products, processes, or services for their whole life cycle. The PSILCA covers 189 individual countries represented by around 15,000 units classified by entities (i.e., industries and commodities). Currently, there are 54 indicators grouped into 18 categories and 4 affected stakeholders, and it is expected to reach 88 indicators [50].

The SHDB database is a project, which was developed by New Earth in 2009 and published in 2013. The project seeks to provide in-depth information on human rights and working conditions along supply chains, to assess risks and provide methods to calculate social footprints. This database covers 113 individual countries represented by around 6500 units classified by entities (i.e., industries and commodities). Currently, there are over 157 indicators grouped into 26 themes and 5 big groups [51].

3. Methodology

Section 2 reviews the most important methodologies used to carry out a complete E-LCA and S-LCA. Although E-LCA is a methodology that is increasingly being implemented, the bibliographic review shows that only a few works have applied E-LCIA methods to evaluate the environmental pillar of sustainability in bridges. These works only use three different E-LCIA methods: CML 2000 (midpoint approach) [52–54], EI99 (endpoint approach) [29], and ReCiPe (midpoint/endpoint approach) [10,31,55,56].

This paper tries to show a methodology to carry out the environmental assessment at all levels. For this purpose, a midpoint/endpoint approach method is necessary. Among the different midpoint/endpoint approaches, the ReCiPe [44,45] method is considered an evolution of CML (midpoint method) and Eco-indicator (endpoint method). The Eco-indicator and CML methods are currently obsolete and have been substituted with the ReCiPe method that has updated characterization and weight factors.

The midpoint approach of the ReCiPe method groups the results into 18 midpoint impact categories, measuring each according to its respective units: agricultural land occupation (ALO), climate change (GWP), fossil depletion (FD), freshwater eutrophication (FEP), freshwater ecotoxicity (FEPT), human toxicity (HTP), ionizing radiation (IRP), marine ecotoxicity (MEPT), marine eutrophication (MEP), particulate matter formation (PMF), metal depletion (MD), natural land transformation (NLT), ozone depletion (OD), photochemical oxidant formation (POFP), terrestrial acidification (TAP), terrestrial ecotoxicity (TEPT), urban land occupation (ULO), and water depletion (WD). These environmental midpoint impact categories have a high level of detail, providing accurate results, although they are more difficult to interpret. The endpoint approach of the ReCiPe method integrates several midpoint impact categories into three endpoint areas of protection: human health (HH), ecosystems (E), and resource availability (R). These endpoint areas of protection have the advantage of being easier to interpret and understand. However, the uncertainty of these results increases due to the high level of aggregation of them. In order to integrate all environmental midpoint impact categories into an overall score, the E-LCIA results are normalized under the use of the ReCiPe normalization factors with respect to Europe per capita emissions [44,57]. In this way, a global score of the total environmental impact caused by the bridge throughout all of its life cycle can be obtained. This overall score is measured in points. In addition, in order to include the long-term perspective of environmental impacts, the hierarchical perspective was used, due to the inclusion of recycling and the subsequent use of steel and concrete for other purposes after the end of the useful life of the structure.

Regarding the S-LCA, although some authors have stated that this methodology is important [3], it is rarely studied, and even less so in the construction sector. The bibliographic review did not find studies that have used the PSILCA or SHDB databases to assess the social pillar of sustainability. This work considers the PSILCA database because it has the most updated available data source, transparent documentation of original data sources, and risk assessment, and provides data quality assessment. In addition, the social information from the PSILCA database can be associated with the processes of the ecoinvent database by the means of an add-on called SOCA developed by Green Delta. In this way, the social assessment can be carried out using the same processes as the environmental assessment, giving coherence to the overall assessment. The SOCA database uses the first version of PSILCA, and provides 54 quantitative and qualitative indicators addressing 18 categories and 4 affected stakeholder groups [48]. In this way, the final indicators of the environmental and social pillars broadly represent the assessment of these pillars, as these indicators group all the information from the databases into the indicators described. In addition, both environmental and social evaluations are carried out using the open source life cycle assessment OpenLCA software. Figure 2 shows the methodology used in this work. To reduce the number of outputs, the endpoint approach of ReCiPe is used to assess the environmental pillar of sustainability, and the indicators provided by the SOCA database are grouped into the four stakeholders represented.

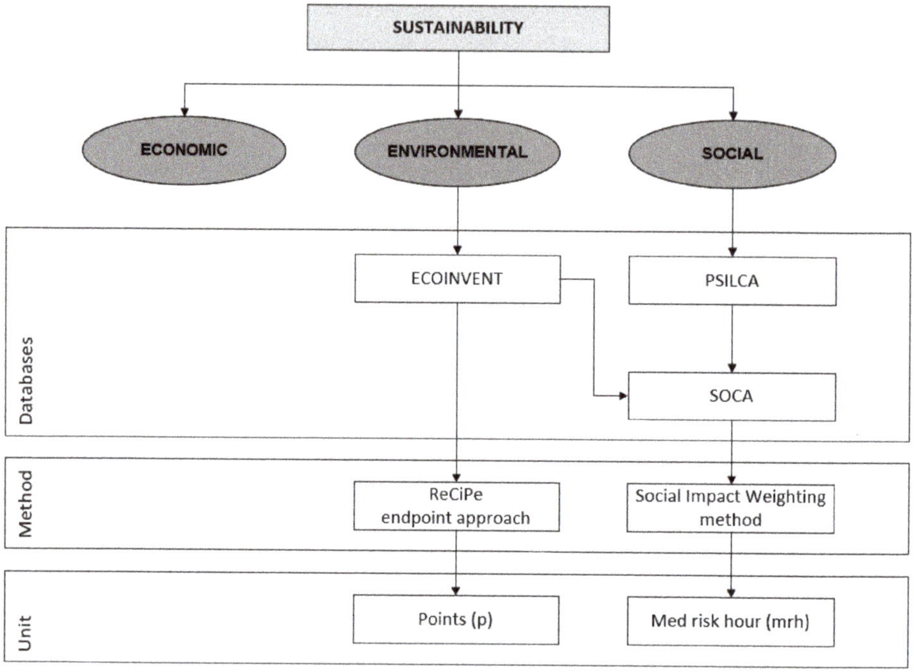

Figure 2. Methodology used in this work.

4. Case Study

4.1. Goal and Scope

The present work aims to provide a comparative life cycle assessment of three alternative concrete bridge designs, from the social and the environmental perspectives. The results of such an approach shall provide valuable information regarding the relationship existing between both dimensions when it comes to the sustainable design of infrastructures. In particular, special emphasis is put on the assessment of the social impacts derived from the different designs under analysis. The conclusions drawn from this case study aim to contribute to the existing knowledge on the social consequences of transport infrastructures.

4.1.1. Functional Unit

Three optimized bridges are analyzed: two box-section post-tensioned concrete road bridges that have different initial and maintenance characteristics, and a pre-stressed concrete precast bridge. These bridges have a width of 12 m and are situated in a seaside region of eastern Spain, whose environment is classified as XC-4 according to EN 206-1 [58]. Therefore, corrosion is mostly due to carbonation and these bridges are subject to the same environmental and traffic conditions. In addition, they have the same width and similar lengths. Therefore, the bridges can be considered equivalent.

The box-section post-tensioned concrete road bridges have a continuous span of 35.2 m, 44 m, and 35.2 m. The first bridge (or alternative A1) was constructed using 50 MPa of concrete and requires one maintenance period, while the second bridge (or alternative A2) was constructed using 35 MPa of concrete and requires two maintenance periods. These bridges are optimized to meet the codes during a service life of 150 years. The distances considered for these bridges are 20 km to carry the aggregate to the concrete plant, 10 km to carry the cement to the concrete plant, 20 km to carry the concrete to the construction place, and 100 km to carry the steel to the construction place.

The pre-stressed concrete precast bridge has three spans of 40 m. This bridge (or alternative A3) was constructed using 35 MPa of concrete in the beams and 40 MPa of concrete in the slab, and requires one maintenance period. This bridge is optimized to meet the codes during a service life of 120 years. The distances considered are 50 km to carry the aggregate to the precast concrete plant, 10 km to carry the cement to the concrete plant, 50 km to carry the precast concrete beams to the construction place, and 100 km to carry the steel to the construction place. Figure 3 shows the three alternatives considered. Due to the total length and service life being a bit different among the three alternatives, the functional unit considered is meter length × year [30,59].

Figure 3. Alternatives.

4.1.2. System Boundaries

The present study considers a "cradle-to-grave" approach, including every relevant manufacturing process related to both the construction and the maintenance activities of each alternative. Given that this assessment is intended for comparative purposes, processes assumed to be identical between alternatives are excluded from the evaluation, following the cut-off criterion established in ISO 14044. This approach has been widely considered in recent comparative life cycle assessments [10,60].

The environmental pillar of the two box-section post-tensioned concrete road bridges have already been assessed by Penadés-Plà et al. [55], and the pre-stressed concrete precast bridge was evaluated by Penadés-Plà et al. [56]. These previous works show the flowchart of the different processes considered in this study. In this work, the social pillar is also considered to obtain a complete life cycle assessment.

4.2. Inventory Analysis

Table 3 shows the quantity of material per 1 m^2 of bridge and the dosage required to manufacture 1 m^3 of concrete according to the concrete strength level. The concrete manufacturing residues are as indicated by Marceau et al. [61]. Reinforced steel is achieved as a mix of the different methods of steel production in accordance with the place of the study. In Spain, the electric arc furnace process manufactures about 67% of steel, while the basic oxygen furnace process manufactures the other 33% of steel. Taking the same steel recycled ratio for each process as in ecoinvent (100% in the electric arc furnace process and 19% in the basic oxygen furnace process), the steel recycled ratio obtained is 71%. These amounts of materials have been obtained from the design of the bridges which follow the Spanish codes for this structure type [62,63], as also the Eurocodes [64,65]. The serviceability and ultimate limit states of compression and tension stress, punching shear, vertical shear, longitudinal shear, torsion, torsion combined with bending and shear, bending, vibration, and cracking have been checked. Furthermore, the geometrical and constructability requirements have been verified.

The construction is organized in this manner: the concrete box girder road bridges prestressed with post-tensioning tendons and the slab of the pre-stressed concrete precast bridge is supposed to be cast in place. Afterward, the beams of the pre-stressed concrete precast bridge are transported to the construction site with special transport and lifted and positioned using tower cranes. Furthermore, the heavy machinery taken into consideration in this section were classified into two categories: the concrete machinery and post-tensioned steel handlings. The quantity of CO_2 emissions and energy for each category was taken from the BEDEC database [66]. In addition, the concrete needs to be handled by heavy equipment that generates 32.24 kg of CO_2 and requires 123.42 MJ of energy per m^3 of concrete produced. Additionally, the machinery for the production of active reinforcement emits 2.62 kg of CO_2 and consumes 10.2 MJ of energy for each kg of active steel. Finally, the formwork taken into account in this study is made of wood and reusable thrice.

Table 3. Amount of materials.

	A1	A2	A3	
			Precast Concrete Beam	Concrete Slab
Strength (MPa)	50	35	35	40
Passive steel (kg/m^2)	74.67	66.89	12.52	23.92
Active steel (kg/m^2)	19.8	21.98	10.53	–
Concrete (m^3/m^2)	0.67	0.674	0.1117	0.1797
Cement (kg/m^3)	400	300	300	320
Gravel (kg/m^3)	726	848	848	829
Sand (kg/m^3)	1136	1088	1088	1102
Water (kg/m^3)	160	160	160	162
Superplasticizer (kg/m^3)	7	4	4	5

For each rehabilitation period, maintenance interventions and infrastructure closures are the same. Hence, the different strategies differ in the number of maintenance periods necessary. Each period of rehabilitation, whose duration is 7 days, involves the removal of the deteriorated concrete surface and its substitution with repair mortar. Furthermore, the traffic detour was quantified considering the percentage of trucks (12%) of the average daily traffic, equal to 8500 vehicles/day, and computing a detour distance of 2.9 km. The process of concrete rehabilitation consists of several phases. First, the deteriorated concrete cover is removed through water blasting. Secondly, by applying an adhesion coating, an appropriate surface for the correct adherence of the new concrete cover is obtained. To conclude, the concrete cover is built by casting the repair mortar. The aforementioned activities are performed by employing a truck-mounted platform [67]. As explained above, the estimation of energy and CO_2 emissions associated with the use of the machinery was acquired from the BEDEC database [66] and amounts to 584.28 MJ and 46.58 CO_2 for each m^2 repaired per maintenance period. Lastly, fixed CO_2 during the entire service life is taken into consideration [68].

The end of life includes the equipment used for the demolition of the bridge and the management of the materials. In this work, the ratio of recycled steel considered is 71% and all the concrete is crushed and disposed of in a landfill. The crushed concrete is supposed to be completely carbonated, and the ratio of recycled steel considered in the manufacturing phase corresponds to that of the end of life phase. Thus, the life cycle of the bridge is closed.

4.3. *Impact Assessment*

Environmental and social dimensions of sustainability are considered to carry out a complete life cycle assessment: the environmental dimension was evaluated using the ReCiPe method and the ecoinvent database, and the social dimension is assessed by means of the Social Impact Weighting Method and the SOCA database. Due to the large number of indicators in the environmental and social dimensions, this study aims to obtain a smaller number of indicators so that results are understandable and complete for these dimensions. For this purpose, the environmental assessment is made according to the endpoint areas of protection of the endpoint approach of the ReCiPe method, and stakeholders made the social assessment.

The endpoint areas of protection obtained by the endpoint approach of the ReCiPe method that represent the environmental dimension are the ecosystems (E), resources (R), and human health (HH). In addition, the four obtained by the SOCA method that represent the social dimension are workers (W), local communities (LC), society (S), and value chain actors (VCA). Such stakeholders are in accordance with those suggested in the Guidelines [37] and are considered representative of the social context of the Spanish region where the structures under analysis are located.

The weighting step of the impact assessment is considered essential when it comes to the holistic evaluation of the sustainability performance of products. However, subjective weighting may lead to jeopardized solutions that might result in inappropriate solutions [69]. This is particularly relevant when it comes to decision making in the field of sustainability, where the complex relations between criteria are usually in conflict. Consequently, great efforts have been applied in recent times to study the influence of such subjectivity in decision-making processes related to sustainable designs in the field of construction [70–72]. As stated above, the scope of the present work is to draw objective conclusions regarding the environmental and social perspectives of infrastructure design. As a consequence, and given that this study is not intended to provide a decision, but to assess the relations existing between the abovementioned dimensions, the different indicators have been considered equally important. Such an approach has been proved to be consistent when assessing the social impacts related to the design of concrete bridge decks [8]. In this way, the subjective assignment of weights is avoided.

5. Results and Discussion

Tables 4–6 show the environmental and social impacts on sustainability for the alternatives A1, A2, and A3, respectively. These tables show the environmental and social impact for each life cycle stage. Here, impacts have been grouped into four stages: impacts related to the manufacturing of the materials required for the construction of the alternatives, including every extraction activity of raw materials and production of the final construction materials, as well as the transport activities from the production facilities to the installation site. Impact results included under the "construction" category consider those related to the machinery involved in construction activities and in the production of auxiliary elements, such as formwork panels. The impacts related to the production of construction materials, as well as to the energy consumption and transport associated with construction activities related to maintenance have been summarized as "use and maintenance". Finally, the results under "EoL" include the impacts associated with the recycling of materials.

Table 4. Sustainability assessment of A1.

Assessment		Unit	Manufacturing	Construction	Use and Maintenance	EoL	Total
Environmental	HH	p	1.33	0.30	0.42	−0.20	1.86
	R	p	1.05	0.10	0.36	0.02	1.53
	E	p	0.69	0.26	0.18	−0.13	1.01
						Total	4.40
Social	W	mrh	227.17	20.27	57.87	2.25	307.56
	LC	mrh	273.58	22.03	71.49	2.54	369.65
	S	mrh	320.67	25.05	79.56	2.98	428.26
	VCA	mrh	199.67	14.09	56.44	1.90	272.11
						Total	1377.58

Note: p—points, mrh—med risk hour, E—ecosystems, R—resources, HH—human health, W—workers, LC—local communities, S—society, VCA—value chain actors, EoL—impacts associated with the recycling of materials.

Table 5. Sustainability assessment of A2.

Assessment		Unit	Manufacturing	Construction	Use and Maintenance	EoL	Total
Environmental	HH	p	1.13	0.32	0.86	−0.15	2.16
	R	p	0.93	0.11	0.72	0.01	1.77
	E	p	0.57	0.28	0.31	−0.10	1.06
						Total	4.98
Social	W	mrh	197.63	20.68	115.75	2.26	336.31
	LC	mrh	238.77	22.36	142.98	2.55	406.67
	S	mrh	285.49	25.42	159.12	3.00	473.02
	VCA	mrh	174.01	14.34	112.87	1.91	303.14
						Total	1519.14

Note: p—points, mrh—med risk hour, E—ecosystems, R—resources, HH—human health, W—workers, LC—local communities, S—society, VCA—value chain actors, EoL—impacts associated with the recycling of materials.

Table 6. Sustainability assessment of A3.

Assessment		Unit	Manufacturing	Construction	Use and Maintenance	EoL	Total
Environmental	HH	p	0.74	0.08	0.56	0.00	1.38
	R	p	0.64	0.05	0.46	0.03	1.17
	E	p	0.36	0.04	0.23	−0.01	0.63
						Total	3.19
Social	W	mrh	124.88	4.02	82.27	2.81	213.97
	LC	mrh	151.20	5.08	101.62	3.44	261.34
	S	mrh	182.92	5.93	113.08	4.05	305.98
	VCA	mrh	109.66	4.11	80.22	2.73	196.72
						Total	978.02

Note: p—points, mrh—med risk hour, E—ecosystems, R—resources, HH—human health, W—workers, LC—local communities, S—society, VCA—value chain actors, EoL—impacts associated with the recycling of materials.

In general, the manufacturing phase is the life cycle stage with the highest impact in every alternative. A3 has the lowest impact for all the indicators. However, A1 has a lower impact in the use and maintenance and end-of-life phases. This is because A1 requires one maintenance period for 150 years of service life, while A2 requires two maintenance periods for the same service life and A3 requires one maintenance period for 120 years of service life. Therefore, A1 has the lowest ratio between maintenance days and service life. Similar trends were recently observed by Tait and Cheung [73] and García-Segura et al. [71], who concluded that, in general, sustainable solutions based on the use of conventional construction materials should focus on reducing the maintenance needs of the designs, given the relevance of this stage in aggressive environments. It is interesting to note that the greater the surface of the deck exposed to chlorides, the greater the impact related to each maintenance activity from both the environmental and the social perspectives.

Figure 4 compares the social and environmental impacts of the three alternatives for each life cycle stage. For this purpose, the upper vertical axis represents the social impact, and the lower vertical axis represents the environmental impact. It is observed that, in general, the negative social impacts considered here are proportional to the environmental impacts associated with each alternative and, consequently, there is a symmetry between these two dimensions of sustainability. However, this observation shall be considered carefully, as it is highly dependent on the social context associated with each alternative analyzed [74]. The observed proportionality is due to the fact that the three alternatives under study are assumed to affect the same social system.

A3 has the lowest global social and environmental impacts and the lowest social and environmental impacts in the manufacturing and construction stages. However, A1 has the lowest social and environmental impacts in the use and maintenance and end-of-life phase. The manufacturing stage has the highest contribution to both impacts. It is observed that, but for the alternative A1, the impacts related to maintenance take a significant proportion of the total life cycle impacts both from a social as well as from an environmental perspective. Such results are in good accordance with other studies on the life cycle impacts of bridges. Navarro et al. [75] observed that, in chloride-laden environments, environmental impacts related to maintenance can even double the impacts related to construction in very aggressive exposures.

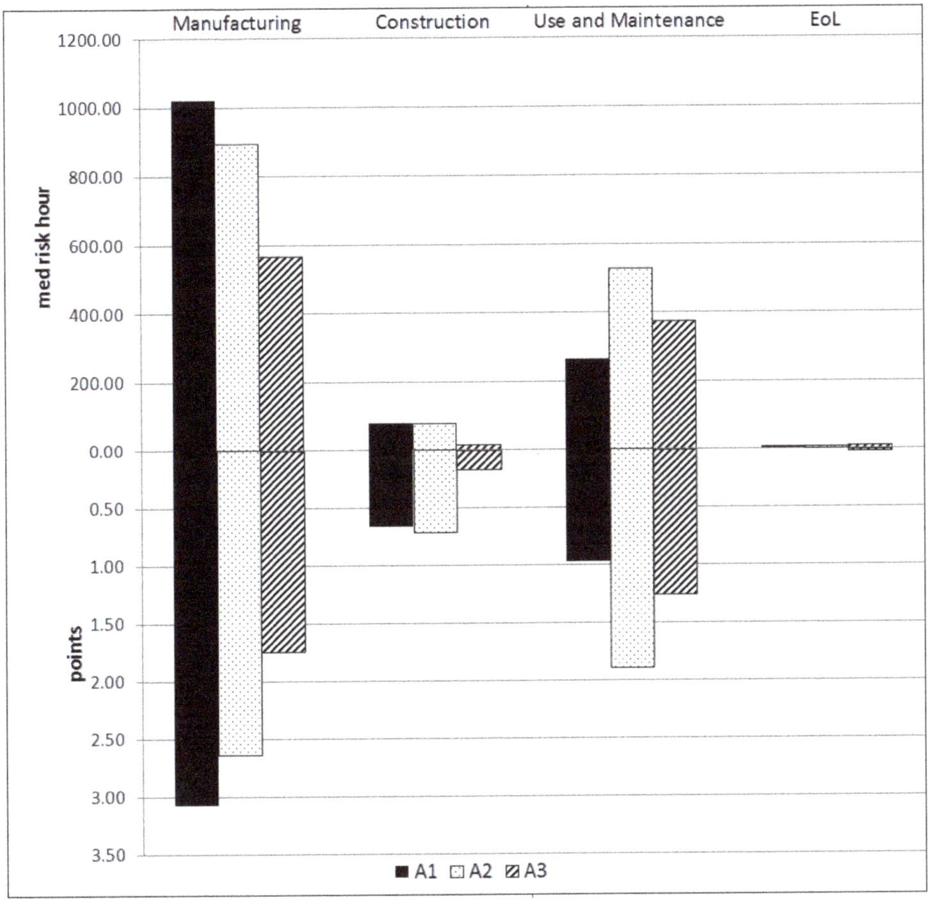

Figure 4. Total social and environmental impact.

In addition, more detailed information can be obtained for each social indicator. Thirteen social indicators have been selected according to a hot spot analysis carried out to identify the relevant social concerns for the specific location of the case study analyzed [76]: association and bargaining rights (ACB), non-fatal accidents (NFA), fatal accidents (FA), gender wage gap (GW), violations of employment laws and regulations (VL), safety measures (SM), frequency of forced labor (FL), trade unionism (TU), fair salary (FS), workers affected by natural disasters (ND), weekly hours of work per employee (WH), social security expenditures (SS) and international migrant workers (IMW). Table 7 shows the influence of the main materials used along the whole bridge life cycle for the selected social indicators. Both concrete and steel manufacturing are the processes with the biggest impacts. This table shows that steel production is the bridge process with the main social impact, followed by concrete production. However, there are two indicators for which diesel consumption has the highest contribution: FA and IMW. Figures 5 and 6 show the contribution of steel production, concrete production, and diesel consumption in these indicators. These figures show that the contribution of diesel consumption in A1 is relatively weak when compared with A2 and A3 because the importance of the materials is higher for A1. In A2 and A3, around half of the impact is due to diesel consumption.

Table 7. Contribution of material in the social impact.

	A1		A2		A3	
	Steel	**Concrete**	**Steel**	**Concrete**	**Steel**	**Concrete**
FL	57.93%	24.11%	53.04%	17.82%	54.82%	16.46%
FS	55.63%	29.60%	53.14%	22.73%	55.42%	21.10%
WH	49.90%	27.02%	43.92%	19.26%	45.12%	17.74%
GW	41.82%	42.58%	42.18%	34.56%	48.36%	34.60%
NFA	31.87%	49.36%	30.03%	37.22%	31.67%	34.12%
FA	35.70%	28.26%	28.27%	18.23%	28.64%	16.59%
SM	24.51%	49.95%	23.15%	36.30%	26.02%	36.09%
ND	51.76%	28.34%	46.35%	20.56%	47.38%	18.65%
SS	50.01%	27.31%	44.01%	19.49%	45.21%	17.88%
VL	53.19%	29.07%	49.35%	21.92%	51.51%	20.32%
ACB	58.06%	29.96%	57.53%	23.94%	62.05%	22.52%
TU	48.20%	29.57%	42.91%	21.32%	44.28%	19.64%
IMW	37.90%	22.07%	28.26%	13.29%	27.11%	11.98%

Note: FL—frequency of forced labor, FS—fair salary, WH—weekly hours of work per employee, GW—gender wage gap, NFA—non-fatal accidents, FA—fatal accidents, SM—safety measures, ND—workers affected by natural disasters, SS—social security expenditures, VL—violations of employment laws and regulations, ACB—association and bargaining rights, TU—trade unionism, IMW—international migrant workers.

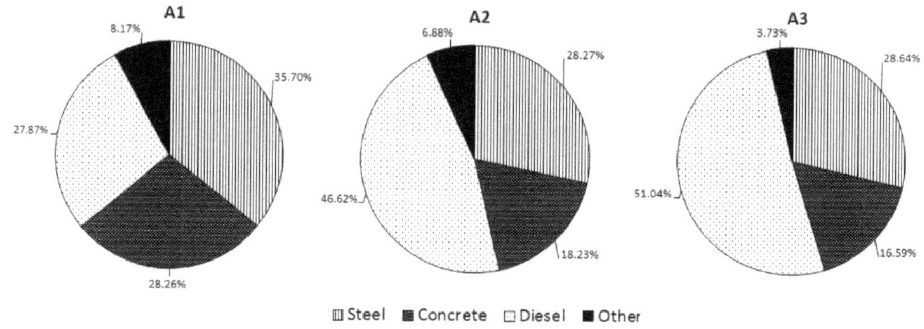

Figure 5. Contribution of processes to FA social impact.

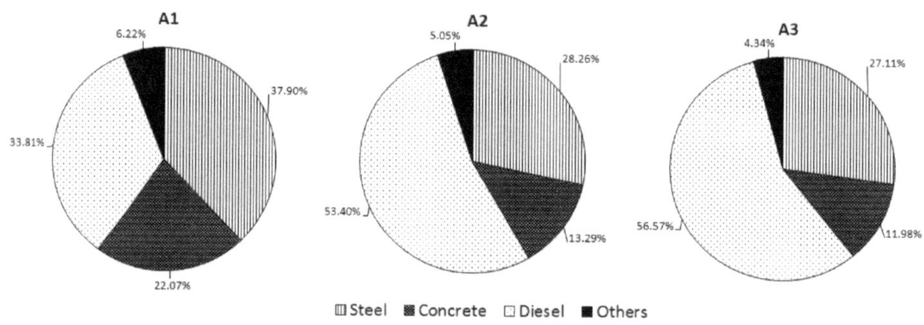

Figure 6. Contribution of processes to IMW social impact.

6. Conclusions

This work carried out a complete life cycle assessment of three bridges: two box-section post-tensioned concrete road bridges, with different initial and maintenance characteristics, and a pre-stressed concrete precast bridge. For this purpose, the environmental and social pillars were evaluated following the LCA methodology. After reviewing and discussing the different LCIA that best represent these pillars, a full sustainability assessment was performed using existing LCIA methods. The ReCiPe method and ecoinvent database were used to carry out the environmental assessment, and the Social Impact Weighting Method and the PSILCA database with the SOCA add-on were used to carry out the social assessment.

The comparison between the three bridges shows that the most sustainable bridge is the pre-stressed concrete precast bridge. This bridge has the lowest impact for all environmental and social indicators. In addition, when the different phases of the bridge life cycle are compared, results show that the manufacturing stage has the highest environmental and social impact. In this phase, concrete production is the process with the highest environmental impact, and steel production is the process with the highest social impact. Focusing on the social assessment, the processes of concrete and steel production have a higher contribution to the social impact. However, other indicators such as the FA and INW are more affected by diesel consumption.

This work aims to propose a complete methodology to evaluate the environmental and social sustainability of bridges using a small number of indicators. This methodology can be applied to other case studies. However, this study has potential limitations. One limitation is that results cannot be compared with other works as they use different methodologies. The wide variety of methods means that the contrast of the environmental and social assessment can only be done with works that use the same methodology, and this leads to a global loss of information about the sustainability assessment. For this reason, future research may unify the methodology to carry out environmental and social assessments. Thus, engineers would have a standard methodology to choose the most sustainable structure among different alternatives.

Author Contributions: This paper represents a result of teamwork. The authors jointly designed the research. V.P.-P. drafted the manuscript. D.M.-M., T.G.-S., I.J.N., and V.Y. edited and improved the manuscript until all authors were satisfied with the final version. All authors have read and agreed to the published version of the manuscript.

Funding: This research was funded by the Ministerio de Economía, Ciencia y Competitividad and FEDER funding grant number [BIA2017-85098-R].

Conflicts of Interest: The authors declare no conflict of interest.

References

1. United Nations. *World Commission on Environment and Development Our Common Future*; Oxford University Press: New York, NY, USA, 1987.
2. United Nations. *Transforming our world: The 2030 agenda for sustainable development*; General Assembly: New York, NY, USA, 2015.
3. Murphy, K. The social pillar of sustainable development: A literature review and framework for policy analysis. *Sustain. Sci. Pract. Policy* **2012**, *8*, 15–29. [CrossRef]
4. Montalbán-Domingo, L.; García-Segura, T.; Sanz, M.A.; Pellicer, E. Social sustainability criteria in public-work procurement: An international perspective. *J. Clean. Prod.* **2018**, *198*, 1355–1371. [CrossRef]
5. Vallance, S.; Perkins, H.C.; Dixon, J.E. What is social sustainability? A clarification of concepts. *Geoforum* **2011**, *42*, 342–348. [CrossRef]
6. Harris, J.M.; Wise, T.A.; Gallagher, K.P.; Goodwine, N.R. *A Survey of Sustainable Development: Social and Economic Dimension*; Island Press: Washinton, DC, USA, 2001.
7. Sierra, L.A.; Yepes, V.; Pellicer, E. Assessing the social sustainability contribution of an infrastructure project under conditions of uncertainty. *Environ. Impact Assess. Rev.* **2017**, *67*, 61–72. [CrossRef]
8. Navarro, I.J.; Yepes, V.; Martí, J.V. Social life cycle assessment of concrete bridge decks exposed to aggressive environments. *Environ. Impact Assess. Rev.* **2018**, *72*, 50–63. [CrossRef]
9. Navarro, I.J.; Martí, J.V.; Yepes, V. Reliability-based maintenance optimization of corrosion preventive designs under a life cycle perspective. *Environ. Impact Assess. Rev.* **2019**, *74*, 23–34. [CrossRef]

10. Navarro, I.J.; Yepes, V.; Martí, J.V. Sustainability assessment of concrete bridge deck designs in coastal environments using neutrosophic criteria weights. *Struct. Infrastruct. Eng.* **2020**, *6*, 949–967. [CrossRef]
11. Montalbán-Domingo, L.; García-Segura, T.; Amalia Sanz, M.; Pellicer, E. Social Sustainability in Delivery and Procurement of Public Construction Contracts. *J. Manag. Eng.* **2019**, *35*, 1–11. [CrossRef]
12. Valdes-Vasquez, R.; Klotz, L.E. Social sustainability considerations during planning and design: Framework of processes for construction projects. *J. Constr. Eng. Manag.* **2013**, *139*, 80–89. [CrossRef]
13. Almahmoud, E.; Doloi, H.K. Assessment of social sustainability in construction projects using social network analysis. *Facilities* **2015**, *30*, 152–176. [CrossRef]
14. Sánchez-Garrido, A.J.; Yepes, V. Multi-criteria assessment of alternative sustainable structures for a self-promoted, single-family home. *J. Clean. Prod.* **2020**, *258*, 120556. [CrossRef]
15. Navarro, I.; Yepes, V.; Martí, J. Life cycle cost assessment of preventive strategies applied to prestressed concrete bridges exposed to chlorides. *Sustainability* **2018**, *10*, 845. [CrossRef]
16. Salas, J.; Yepes, V. VisualUVAM: A decision support system addressing the curse of dimensionality for the multi-scale assessment of urban vulnerability in Spain. *Sustainability* **2019**, *11*, 2191. [CrossRef]
17. Kripka, M.; Yepes, V.; Milani, C.J. Selection of sustainable short-span bridge design in Brazil. *Sustainability* **2019**, *11*, 1307. [CrossRef]
18. Penadés-Plà, V.; Yepes, V.; García-Segura, T. Robust decision-making design for sustainable pedestrian concrete bridges. *Eng. Struct.* **2020**, *209*, 109968. [CrossRef]
19. Rasheed, A.; Farooq, S.H.; Usman, M.; Hanif, A.; Khan, N.A.; Khushnood, R.A. Structural reliability analysis of superstructure of highway bridges on China-Pakistan Economic Corridor (CPEC): A case study. *J. Struct. Integr. Maint.* **2018**, *3*, 197–207. [CrossRef]
20. Hansen, M. Determination and assessment of fatigue stresses on concrete bridges. *Struct. Concr.* **2020**, 1–12. [CrossRef]
21. Sargsyan, A.; Sargsyan, G.; Resnik, B. Influence of Cracks on Frequency of the Self-Vibration of Reinforced Concrete T Beam of Road Bridges | Scientific.Net. *Key Eng. Mater.* **2019**, *828*, 9–13. [CrossRef]
22. Penadés-Plà, V.; García-Segura, T.; Martí, J.; Yepes, V. A review of multi-criteria decision-making methods applied to the sustainable bridge design. *Sustainability* **2016**, *8*, 1295. [CrossRef]
23. Horvath, A.; Hendrickson, C. Steel versus steel-reinforced concrete bridges: Environmental assessment. *J. Infrastruct. Syst.* **1998**, *4*, 111–117. [CrossRef]
24. Widman, J. Environmental impact assessment of steel bridges. *J. Constr. Steel Res.* **1998**, *46*, 291–293. [CrossRef]
25. Stengel, T.; Schiessl, P. Life cycle assessment of UHPC bridge constructions: Sherbrooke Footbridge, Kassel Gärtnerplatz Footbridge and Wapello Road Bridge. *Archit. Civ. Eng. Environ.* **2009**, *1*, 109–118.
26. Gervasio, H.; Simoes da Silva, L. Comparative life-cycle analysis of steel-concrete composite bridges. *Struct. Infrastruct. Eng.* **2008**, *4*, 251–269. [CrossRef]
27. Itoh, Y.; Kitagawa, T. Using CO_2 emission quantities in bridge lifecycle analysis. *Eng. Struct.* **2003**, *25*, 565–577. [CrossRef]
28. Bouhaya, L.; Roy, R.L.; Feraille-Fresnet, A. Simplified environmental study on innovative bridge structure. *Environ. Sci. Technol.* **2009**, *43*, 2066–2071. [CrossRef] [PubMed]
29. Steele, K.N.P.; Cole, G.; Parke, G. Application of life cycle assessment technique in the investigation of brick arch highway bridges. In Proceedings of the 6th International Masonry Conference, London, UK, 9 November 2002; pp. 1–8.
30. Pang, B.; Yang, P.; Wang, Y.; Kendall, A.; Xie, H.; Zhang, Y. Life cycle environmental impact assessment of a bridge with different strengthening schemes. *Int. J. Life Cycle Assess* **2015**, *20*, 1300–1311. [CrossRef]
31. Du, G.; Safi, M.; Pettersson, L.; Karoumi, R. Life cycle assessment as a decision support tool for bridge procurement: Environmental impact comparison among five bridge designs. *Int. J. Life Cycle Assess.* **2014**, *19*, 1948–1964. [CrossRef]
32. Hammervold, J.; Reenaas, M.; Brattebø, H. Environmental life cycle assessment of bridges. *J. Bridg. Eng.* **2013**, *18*, 153–161. [CrossRef]
33. Gervásio, H.; Simões Da Silva, L. A probabilistic decision-making approach for the sustainable assessment of infrastructures. *Expert Syst. Appl.* **2012**, *39*, 7121–7131. [CrossRef]
34. Sabatino, S.; Frangopol, D.M.; Dong, Y. Sustainability-informed maintenance optimization of highway bridges considering multi-attribute utility and risk attitude. *Eng. Struct.* **2015**, *102*, 310–321. [CrossRef]

35. Chen, Z.; Abdullah, A.B.; Anumba, C.J.; Li, H. ANP experiment for demolition plan evaluation. *J. Constr. Eng. Manag.* **2013**, *140*, 51–60. [CrossRef]
36. International Organization for Standardization (ISO). *International Organization for Standardization (ISO) Environmental Managment—Life Cycle Assessment—Principles and Framework*; International Organization for Standarization: Geneva, Switzerland, 2006.
37. Benoît, C.; Mazijn, B. *Guidelines for Social Life Cycle Assessment of Products*; UNEP/SETAC Life Cycle Initiative, Sustainable Product and Consumption Branch: Paris, France, 2009; Volume 15, ISBN 9789280730210.
38. Guinée, J.B.; Gorrée, M.; Heijungs, R.; Huppes, G.; Kleijn, R.; Wegener Sleeswijk, A.; Udo De Haes, H.A.; de Bruijn, J.A.; van Duin, R.; Huijbregts, M.A.J. Life cycle assessment: An operational guide to the ISO standards. *III Sci. Backgr.* **2001**, 692. [CrossRef]
39. Hauschild, M.; Potting, J. Background for spatial differentiation in LCA impact assessment—The EDIP2003 methodology. *Environ. News* **2005**, *80*, 1–195.
40. Bare, J.C. The Tool for the Reduction and Assessment of Chemical and Other Environmental Impacts. *J. Ind. Ecol.* **2002**, *6*, 49–78. [CrossRef]
41. Gocdkoop, M.; Hofstetter, P.; Müller-Wenk, R.; Spriemsma, R. The Eco-Indicator 98 explained. *Int. J. Life Cycle Assess.* **1998**, *3*, 352–360. [CrossRef]
42. Steen, B. *A Systematic Approach to Environmental Priority Strategies in Product Development (EPS)*; Centre for Environmental Assessment of Products and Material System: Gothenburg, Sweeden, 1999.
43. Frischknecht, R.; Steiner, R.; Jungbluth, N. Methode der ökologischen Knappheit—Ökofaktoren 2006. 2009.
44. Goedkoop, M.; Heijungs, R.; Huijbregts, M.; De Schryver, A.; Struijs, J.; Van Zelm, R. *ReCiPe 2008 A Life Cycle Impact Assessment Method which Comprises Harmonised Category Indicators at the Midpoint and the Endpoint Level*; Ministerie van VROM: Hague, The Netherlands, 2009.
45. Huijbregts, M.A.J.; Steinmann, Z.J.N.; Elshout, P.M.F.; Stam, G.; Verones, F.; Vieira, M.D.M.; Hollander, A.; Zijp, M.; Zelm, R. *ReCiPe 2016 A Harmonized Life Cycle Impact Assessment Method at Midpoint and Endpoint Level Report I: Characterization*; Rijksinstituut voor Volksgezondheid en Milieu RIVM: De Bilt, The Netherlands, 2016.
46. Itsubo, N.; Sakagami, M.; Washida, T.; Kokubu, K.; Inaba, A. Weighting across safeguard subjects for LCIA through the application of conjoint analysis. *Int. J. Life Cycle Assess.* **2004**, *9*, 196–205. [CrossRef]
47. Jolliet, O.; Margni, M.; Charles, R.; Humbert, S.; Payet, J.; Rebitzer, G.; Rosenbaum, R. IMPACT 2002+: A new Life Cycle Impact Assessment Methodology. *Int. J. Life Cycle Assess.* **2003**, *8*, 324–330. [CrossRef]
48. *GreenDelta PSILCA v1.0 (Product Social Impact Life-Cycle Assessment)*; Greendelta: Berlin, Germany, 2013.
49. New Earth SHDB v1.0 (Social Hotspot Database). 2009. Available online: https://www.socialhotspot.org/ (accessed on 1 October 2019).
50. GreenDelta PSILCA Database. Available online: https://psilca.net/ (accessed on 1 October 2019).
51. New Earth SHDB Database. Available online: https://www.socialhotspot.org/for-more-information.html (accessed on 1 October 2019).
52. Du, G.; Karoumi, R. Environmental life cycle assessment comparison between two bridge types: Reinforced concrete bridge and steel composite bridge. In Proceedings of the 3rd International Conference on Sustainable Construction Materials and Technologies, Kyoto, Japan, 18–21 August 2013.
53. Hettinger, A.; Birat, J.; Hechler, O.; Braun, M. Sustainable bridge—LCA for a composite and a concrete bridge. In *Economical Bridge Solutions Based on Innovative Composite Dowels and Integrated Abutments*; Petzek, E., Bancila, R., Eds.; Springer Vieweg: Wiesbaden, Germany, 2015; pp. 45–54.
54. Du, G.; Karoumi, R. Life cycle assessment of a railway bridge: Comparison of two superstructure designs. *Struct. Infrastruct. Eng.* **2012**, *9*, 1149–1160. [CrossRef]
55. Penadés-Plà, V.; Martí, J.V.; García-Segura, T.; Yepes, V. Life-cycle assessment: A comparison between two optimal post-tensioned concrete box-girder road bridges. *Sustainability* **2017**, *9*, 1864. [CrossRef]
56. Penadés-Plà, V.; García-Segura, T.; Martí, J.V.; Yepes, V. An optimization-LCA of a prestressed concrete precast bridge. *Sustainability* **2018**, *10*, 685. [CrossRef]
57. Pons, J.J.; Penadés-Plà, V.; Yepes, V.; Martí, J.V. Life cycle assessment of earth-retaining walls: An environmental comparison. *J. Clean. Prod.* **2018**, *192*, 411–420. [CrossRef]
58. European Committee for Standardization. *EN 206-1 Concrete—Part1: Specification, Performance, Production and Conformity*; European Committee for Standardization: Brussels, Belgium, 2000.
59. Steele, K.; Cole, G.; Parke, G.; Clarke, B.; Harding, J.; Harding, J. Highway bridges and environment-sustainable perspectives. *Proc. Inst. Civ. Eng.* **2003**, *156*, 176–182. [CrossRef]

60. Martínez-Blanco, J.; Lehmann, A.; Muñoz, P.; Antón, A.; Traverso, M.; Rieradevall, J.; Finkbeiner, M. Application challenges for the social Life Cycle Assessment of fertilizers within life cycle sustainability assessment. *J. Clean. Prod.* **2014**, *69*, 34–48. [CrossRef]
61. Marceau, M.L.; Nisbet, M.A.; Vangeem, M.G. *Life Cycle Inventory of Portland Cement Concrete*; Portland Cement Association: Skokie, IL, USA, 2007.
62. Ministerio de Fomento. *EHE-08: Code on Structural Concrete*; Ministerio de Fomento: Madrid, Spain, 2008.
63. Ministerio de Fomento. *IAP-11: Code on the Actions for the Design of Road Bridges*; Ministerio de Fomento: Madrid, Spain, 2011.
64. European Committee for Standardization. *EN 1001-2:2003. Eurocode 1: Actions on Structures—Part 2: Traffic Loads Bridges*; European Committee for Standardization: Brussels, Belgium, 2003.
65. European Committee for Standardisation. *EN1992-2:2005. Eurocode 2: Design of Concrete Structures—Part 2: Concrete Bridge-Design and Detailing Rules*; European Committee for Standardization: Brussels, Belgium, 2005.
66. Catalonia Institute of Construction Technology. BEDEC PR/PCT ITEC Material Database. 2016. Available online: https://metabase.itec.cat/vide/ca/bedec (accessed on 1 October 2019).
67. García-Segura, T.; Yepes, V.; Frangopol, D.M.; Yang, D.Y. Lifetime reliability-based optimization of post-tensioned box-girder bridges. *Eng. Struct.* **2017**, *145*, 381–391. [CrossRef]
68. García-Segura, T.; Yepes, V.; Alcalá, J. Life cycle greenhouse gas emissions of blended cement concrete including carbonation and durability. *Int. J. Life Cycle Assess.* **2014**, *19*, 3–12. [CrossRef]
69. Radwan, N.M.; Senousy, M.B.; El, A.; Riad, D.M. Neutrosophic AHP multi criteria decision making method applied on the selection of learning management system. *Int. J. Adv. Comput. Technol.* **2016**, *8*, 95–105.
70. Prascevic, N.; Prascevic, Z. Application of fuzzy AHP for ranking and selection of alternatives in construction project management. *J. Civ. Eng. Manag.* **2017**, *23*, 1123–1135. [CrossRef]
71. García-Segura, T.; Penadés-Plà, V.; Yepes, V. Sustainable bridge design by metamodel-assisted multi-objective optimization and decision-making under uncertainty. *J. Clean. Prod.* **2018**, *202*, 904–915. [CrossRef]
72. Pamučar, D.; Badi, I.; Sanja, K.; Obradović, R. A novel approach for the selection of power-generation technology using a linguistic neutrosophic CODAS method: A case study in Libya. *Energies* **2018**, *11*, 2489. [CrossRef]
73. Tait, M.W.; Cheung, W.M. A comparative cradle-to-gate life cycle assessment of three concrete mix designs. *Int. J. Life Cycle Assess.* **2016**, *21*, 847–860. [CrossRef]
74. Sierra, L.A.; Pellicer, E.; Yepes, V. Method for estimating the social sustainability of infrastructure projects. *Environ. Impact Assess. Rev.* **2017**, *65*, 41–53. [CrossRef]
75. Navarro, I.J.; Yepes, V.; Martí, J.V.; González-Vidosa, F. Life cycle impact assessment of corrosion preventive designs applied to prestressed concrete bridge decks. *J. Clean. Prod.* **2018**, *196*, 698–713. [CrossRef]
76. Hosseinijou, S.A.; Mansour, S.; Shirazi, M.A. Social life cycle assessment for material selection: A case study of building materials. *Int. J. Life Cycle Assess.* **2014**, *19*, 620–645. [CrossRef]

sustainability

MDPI

Article

Application of Fuzzy and Rough Sets to Environmental Zonation for Concrete Durability: A Case Study of Shaanxi Province, China

Daming Luo [1,2,*], Yan Wang [1,3], Shaohui Zhang [2] and Ditao Niu [1,2]

1 State Key Laboratory of Green Building in Western China, Xi'an University of Architecture and Technology, Xi'an 710055, China; wangyanwjx@126.com (Y.W.); niuditao@163.com (D.N.)
2 School of Civil Engineering, Xi'an University of Architecture and Technology, Xi'an 710055, China; zhangshaohui@xauat.edu.cn
3 College of Materials Science and Engineering, Xi'an University of Architecture and Technology, Xi'an 710055, China
* Correspondence: dmluo@xauat.edu.cn; Tel.: +86-156-6708-8332

Received: 13 March 2020; Accepted: 6 April 2020; Published: 13 April 2020

Abstract: The durability of concrete structures is influenced by various factors, and the durability damage mechanism is different when the structure is in different environmental conditions. This will have implications for improving the durability of concrete structures and extending its service life if the special environmental condition is taken into account in the durability design. Aimed at the environmental zonation for concrete durability, this paper investigated the durability factors influencing concrete structures in Shaanxi Province, China, including atmospheric temperature, precipitation, corrosive gas, and acid rain. The variations of the above-mentioned factors were analyzed and the indexes of environmental zonation were proposed. According to the zoning principle, the weights of zoning indexes calculated using fuzzy rough sets were used to divide Shaanxi Province into three first-level zones, namely the Freeze-Thaw Cycle Zone, the Neutralization–Freeze-Thaw Interaction Zone, and the Neutralization Zone. These three zones were then subdivided into nine second-level zones. The main mechanism of concrete deterioration and the environmental characteristics of all zones were then analyzed. The method proposed in this paper puts forward clear zoning indexes and quantifies them, which can improve the quality and accuracy of the zoning results. Moreover, the research achievements are helpful for engineers to reduce the impact of the environment on structure and the maintenance cost during the structural service life to a certain extent.

Keywords: concrete structure; environmental zonation; concrete durability; fuzzy set; rough set

1. Introduction

The design, construction, operation, and management of concrete structures are affected by its in-service environment to varying degrees [1]. Due to the various environmental conditions of concrete structures, the mechanism of durability deterioration is complex. For instance, carbonization [2], freeze-thaw damage [3–5], acid rain [6,7], and ion erosion [8–12] can directly affect the durability of concrete structures. It is necessary to take into account the in-service environmental condition when it comes to the design of new concrete structures and the evaluation and maintenance of existing concrete structures.

Durability environmental zonation is to divide a country or region into different zones according to the environmental conditions and their influence on the durability of concrete structures. Some durability design codes and standards, such as the *Code for Design of Concrete Structures* (GB 50010-2010) [13], the *Standard for Design of Concrete Structure Durability* (GB/T 50476-2019) [14], and the *Eurocode 2: Design of concrete Structures* (EN 1992-1-1:2004) [15], classify the service environment of concrete structures into different categories. However, some of the above-mentioned codes and most of the studies pay more attention to the measures of concrete durability, while not much attention is paid to the differences in the environment of different regions and the damage degree of the structure under the corrosive environment. In view of this, some scholars have studied the environmental zonation of concrete structures and acquired some achievements [16,17], or the economic impacts of sustainable vertical extension methods for existing underground spaces [18]. However, these studies did not select appropriate zoning indexes nor quantify them in the process of durability zonation. They mainly focused on the qualitative zonation according to the impact of the environment on the durability of concrete structures, which is subjective and insufficient to reflect the actual situation. If the durability design can be carried out according to the local environmental conditions and durability damage forms, the service life of the building can be prolonged, the maintenance cost can be reduced, and the life-cycle construction cost can be saved.

Taking Shaanxi Province in Northwest China for example, the influence of various environmental factors on the durability of concrete structures is analyzed, and the index of environmental zonation is determined on the basis of investigating and analyzing the environmental conditions and durability status of concrete structures. Based on fuzzy rough sets, the influence of environmental factors on structural durability will be obtained, and the environmental zoning map of Shaanxi Province for concrete durability will be achieved.

2. Analysis of Durability Environmental Conditions in Shaanxi Province

Shaanxi Province, located in Northwest China, is a long and narrow region with a varied topography. From north to south, it can be divided into three geomorphic areas: Northern Shaanxi, Guanzhong, and Southern Shaanxi. Southern Shaanxi has a sub-tropical and humid climate, Guanzhong and parts of Northern Shaanxi have a warm temperate and semi-humid climate, and Northern Shaanxi has a mid-temperate and semi-arid climate along the Great Wall [19].

The environmental factors affecting the durability of concrete structures in Shaanxi Province can be divided into two categories: (1) climatic conditions, namely atmospheric temperature, atmospheric humidity, etc., which are closely related to concrete carbonization and the freeze-thaw cycle; (2) erosion medium, including various corrosive gases in the atmosphere and ions dissolved in water. With the rapid economic development of Shaanxi Province, the acid rain is becoming more frequent and may be the most important erosion medium during the service period of the concrete structures.

This paper selected meteorological data of 93 counties and cities in Shaanxi Province, among which, 28 are in Southern Shaanxi, 40 in Guanzhong, and 25 in Northern Shaanxi. For the missing data of some counties and cities, the data of the surrounding counties and cities with similar environmental conditions were used. The developing trend of ambient temperature, relative humidity, corrosive gases, and acid rain were analyzed.

2.1. Atmospheric Temperature

2.1.1. Annual Average Temperature

The annual average temperature of Shaanxi Province is 13.7 °C, varying from south to north and east to west: 7~12 °C in Northern Shaanxi, 12~14 °C in Guanzhong, and 14~16 °C in Southern Shaanxi. In the last half century, the average annual temperature in Northern Shaanxi, Guanzhong, and Southern Shaanxi have shown a fluctuating upward trend (Figure 1), and its variation coefficient is between 0.03 and 0.07 (Table 1) [20]. For Northern Shaanxi, the annual average temperature peaked in 1998 (10.63 °C), and the minimum value is in 1967 (7.97 °C). For Guanzhong, the annual average temperature reached the highest value (13.31 °C) in 2016 and the lowest value (11.13 °C) in 1984. For Southern Shaanxi, the annual average temperature showed its highest value (14.61 °C) in 1998, and its lowest value (12.25 °C) in 1967. The variations of the annual average temperature of Northern

Shaanxi, Guanzhong, and Southern Shaanxi were highest in the early 21st century, while Northern Shaanxi reached the lowest in the 1960s and the other two regions hit a record low in the 1980s.

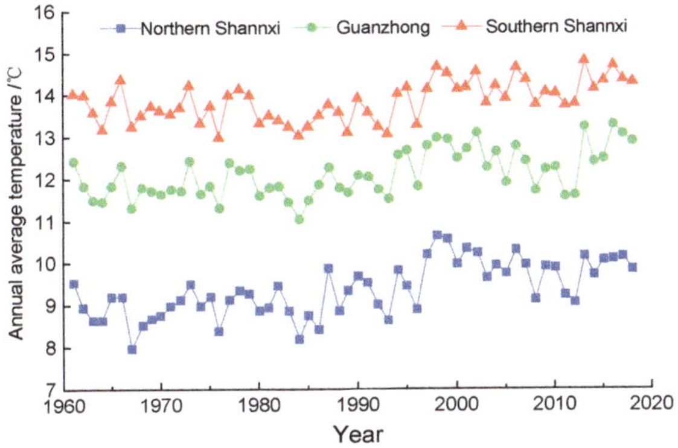

Figure 1. Annual average temperature in Shaanxi Province in the last half century.

Table 1. Statistical characteristics of annual average temperature in Shaanxi Province (°C).

Region	Mean Value	Standard Deviation	Skewness Coefficient	Kurtosis Coefficient	Range of Variation	Range	Variation Coefficient
Northern Shaanxi	9.24	0.62	0.374	−0.245	7.97~10.63	2.66	0.07
Guanzhong	12.08	0.49	0.383	−0.633	11.13~13.15	2.02	0.04
Southern Shaanxi	13.70	0.44	0.183	−0.807	12.97~14.61	1.64	0.03

In order to present the regional distribution of annual average temperature, the annual average temperature is divided into 4 levels, as shown in Figure 2. It can be seen that the annual average temperature of most counties and districts in Northern Shaanxi is below 10 °C, and that of Guanzhong and Southern Shaanxi is 12~14 °C and above 12 °C, respectively.

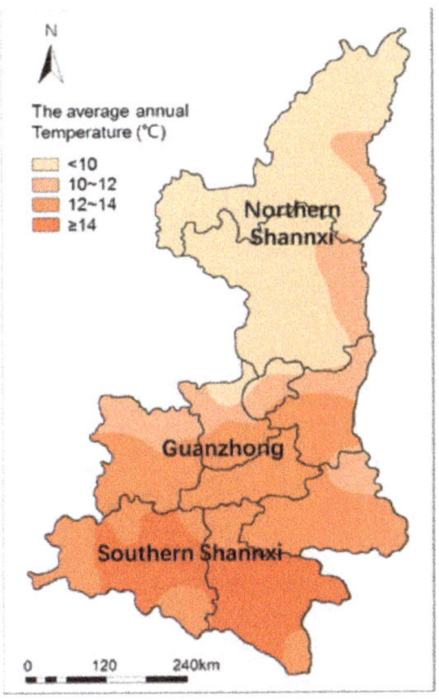

Figure 2. Annual average temperature of Shaanxi Province.

2.1.2. Average Monthly Temperature

As can be seen from the monthly average temperature distribution in the three regions of Shaanxi Province (Figure 3) [20]. The average temperature in January, February, and December in Northern Shaanxi was below 0 °C, while that in other months was above 0 °C and it is the highest in July reaching 22.28 °C. For Guanzhong, the average monthly temperature in January and December was below 0 °C, and that in July was the highest (24.88 °C), followed by June (23.18 °C) and August (23.58 °C). The average temperature in Southern Shaanxi is above 0 °C every month, with the lowest value in January (1.83 °C) and the highest value in July (25.01 °C). On the whole, the temperature changes in Northern Shaanxi, Guanzhong, and Southern Shaanxi are generally the same.

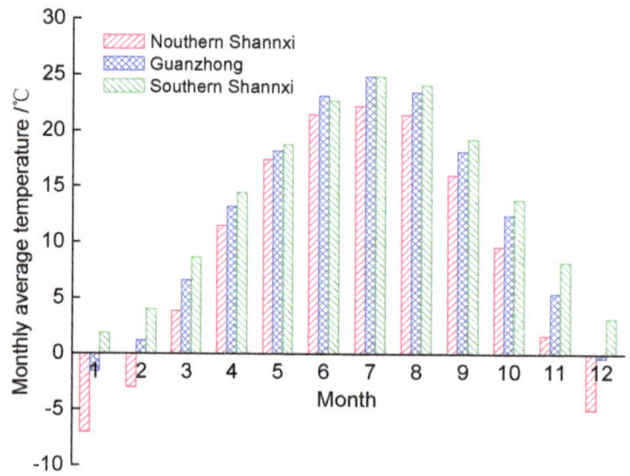

Figure 3. Average monthly temperature distribution in Shaanxi Province.

2.2. *Precipitation*

Shaanxi Province is located in mid-latitude, and the Qinling mountains traverse the south-central part of it. The climate in the south of Qinling mountains is humid, while that in the north is dry. The annual precipitation in Shaanxi has a significant regional characteristic, and it shows a fluctuating downward trend in the three regions (Figure 4) [20]. In Northern Shaanxi, the annual precipitation fluctuated greatly in 1960s, with the maximum value (746.3 mm) in 1964 and the minimum value (293.7 mm) in 1965. In Guanzhong and Southern Shaanxi, the annual precipitation fluctuated greatly from the 1970s to the 1980s, and reached its maximum value in 1983 (Guanzhong: 899.7 mm; South Shaanxi: 1273.7 mm) and the minimum value in 1997 (Guanzhong: 360.3 mm; Southern Shaanxi: 577.6 mm). In general, all regions suffered a downward trend in annual precipitation, with Southern Shaanxi experiencing the largest decline and Northern Shaanxi the smallest. The average decline of the whole province is 2.08 mm·a^{-1}.

In order to demonstrate the spatial distribution of precipitation in Shaanxi Province, the annual precipitation was divided into three levels: 200~400 mm, 400~800 mm, and 800~1600 mm. As can be seen from Figure 5, the average annual precipitation in Southern Shaanxi is more than 800 mm, while that in Guanzhong and Northern Shaanxi is mostly between 400 and 800 mm, and in some counties and cities in Northern Shaanxi is less than 400 mm. In general, the annual average precipitation in Shaanxi Province varies from place to place, with more in the south and less in the north. This difference is mainly related to the topography and geomorphology of Shaanxi Province. The water vapor in the atmosphere of Shaanxi Province mainly comes from the Western Pacific Ocean and the Bay of Bengal, and it is blocked by the Qinling mountains, resulting in strong precipitation in its southern slope. Therefore, the precipitation in Southern Shaanxi is higher than that in Guanzhong and Northern Shaanxi.

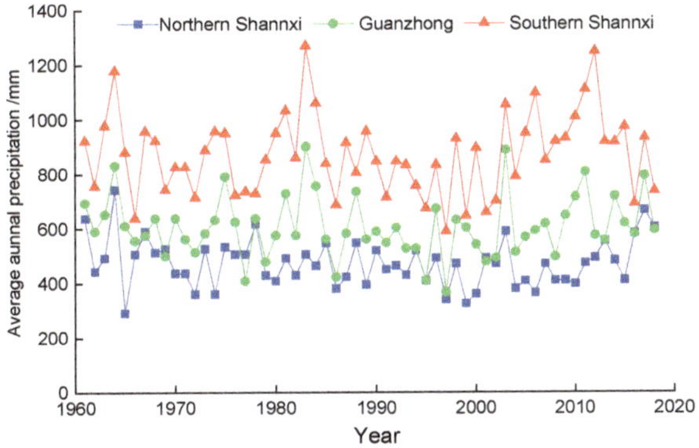

Figure 4. Annual average precipitation in Shaanxi Province from 1961 to 2016.

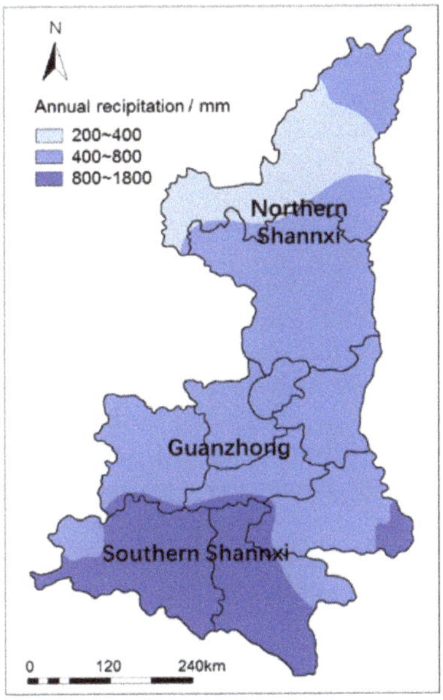

Figure 5. Average annual precipitation in Shaanxi Province.

2.3. *Corrosive Gases in the Atmosphere*

The investigation shows that atmospheric pollution in Shaanxi Province is mainly soot pollution. The exhaust gas from fuel combustion and production processes accounts for 74 percent and 20 percent of the total industrial exhaust gas respectively. The major pollutants in Shaanxi are sulfur dioxide, smoke, nitrogen oxides, and carbon monoxide, which tend to dissolve in water to form acid rain and affects the durability of buildings [21].

2.3.1. Carbon Dioxide

CO_2 is one of the main causes of concrete carbonization and steel corrosion [22]. Shaanxi Province has witnessed rapid economic development in recent years, and overdependence on fossil energy has led to a continuous increase of carbon emissions. According to Figure 6, the annual growth rate of GDP of Shaanxi Province from 1995 to 2017 was 11.6%, and the carbon dioxide emissions increased year by year, from 7.544×10^7 t in 1995 to 30.212×10^7 t in 2017, with an annual growth rate of 6.8% [21].

The trend of carbon emissions and the growth of Gross Domestic Product (GDP) of Shaanxi are roughly the same. However, due to the high mobility of air and the absorption of vegetation, the atmospheric CO_2 concentration does not increase significantly in spite of the increase of carbon emissions, and the province-wide variation of the CO_2 concentration is relatively small.

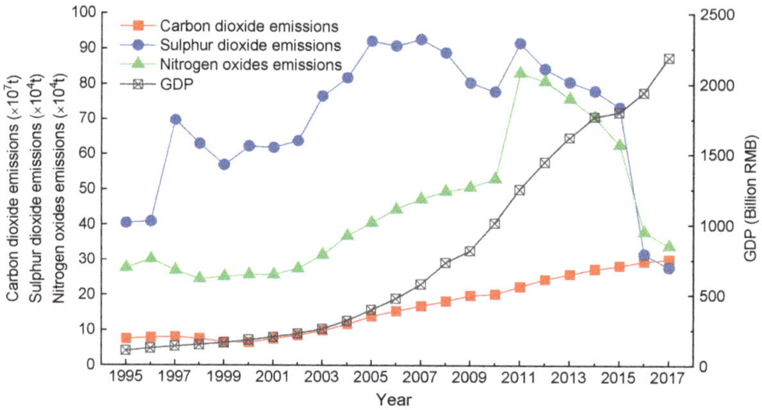

Figure 6. Corrosive gases emissions and economic growth in Shaanxi Province in the last two decades.

2.3.2. Sulfur Dioxide and Nitrogen Dioxide

Sulfur dioxide (SO_2) and nitrogen dioxide (NOx) in the air mainly come from industrial production processes, the burning of fossil fuels, and chemical decomposition in the soil [23,24]. In recent years, the rural industry with large biomass fuel consumption in Shaanxi Province is developing rapidly, while pollution control measures are relatively backward, leading to high corrosive gases emissions. If SO_2 and NOx emissions exceed the environment's capacity, acid rain may occur. Based on the data of the State Environmental Protection Administration shown in Figure 6 [21], the regional difference among the corrosive gas emissions are closely related to economic development and environmental governance. Before 2011, the SO_2 and NOx emissions show an upward trend with a small fluctuation, and this is consistent with the rapid development of the Shaanxi economy. Thanks to the improvement of environmental protection measures and the governance technology, the corrosive gas emissions have been significantly reduced since 2011.

In terms of geographical distribution, the air pollution in large and medium-sized cities in Guanzhong is serious, with average SO_2 emissions over 3.5 t/km^2, among which, Xi'an, Weinan, Baoji, Xianyang, and Tongchuan are the major air pollution areas. The average SO_2 emissions in Northern Shaanxi were 0.3~3.5 t/km^2, and those in Southern Shaanxi were less than 0.3 t/km^2. In addition, Yanchang in Northern Shaanxi is located in the basin valley, so the pollutants are not easy to spread. As for Yulin, the air quality is also poor due to the cold climate and long heating period.

2.4. *Acid Rain*

Precipitation acidity is affected by many factors. In addition to acid substances such as sulfuric acid and nitric acid, soil particles, industrial dust, and natural nitrogen also play an important role [25]. Due to fossil fuel-based energy consumption structure and combustion inefficiency, the acid rain in Shaanxi Province has a high frequency and intensity, and the acid rain distribution has significant regional characteristics: mild in the north and severe in the south, less in the west and more in the east (Figures 7 and 8) [20,21]. The climate in Northern Shaanxi and Guanzhong is dry and rainless, and there are more suspended particles in the air than those in Southern Shaanxi. Meanwhile, the soil particles are neutral or slightly alkaline, which can neutralize the acidic raindrops to some extent.

In contrast, the pH value of precipitation in Southern Shaanxi is relatively low since the acidic soil particles and acid corrosive gas in the air can be easily eluted by rainwater.

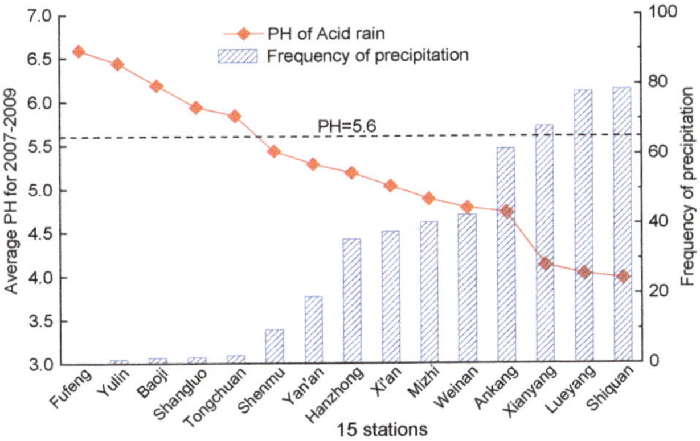

Figure 7. Average pH and frequency of precipitation in 15 stations from 2007 to 2019.

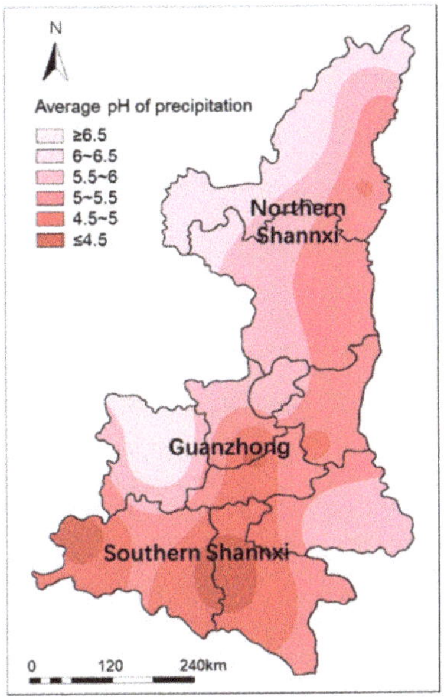

Figure 8. Distribution map of acid rain in Shaanxi Province.

3. Environmental Zonation for Concrete Durability in Shaanxi Province

3.1. Principles of Environmental Zonation for Concrete Durability

The durability design, construction, and management of concrete structures are different due to the different environments in different regions [26–28]. Based on a comprehensive analysis of regional environmental conditions and the durability of concrete structures in Shaanxi Province, the regions with the same or similar climate environments will fall into the same environmental zones. The boundary line of each region is mainly determined by the calculation results, and the road divisions and administrative divisions are also taken into account. The zoning results are convenient to standardize the durability design of concrete structures and to implement the urban construction standards.

3.2. Indexes of Environmental Zonation

The field test results show that the durability damage of concrete structures in Shaanxi Province is mainly caused by concrete neutralization or freeze-thaw damage, and the external environmental factors affecting the durability of concrete mainly include temperature, relative humidity, CO_2 concentration, and acid rain.

Although the concentration of CO_2 in Shaanxi Province has been increasing in recent years, its concentration gradient is small in the whole province [21], so the CO_2 concentration will not be selected as the zoning index. The ambient temperature and relative humidity vary greatly in different regions, resulting in different climatic characteristics in Shaanxi Province [21]. Moreover, the relative humidity has a positive correlation with the precipitation, so the annual precipitation will be selected as the zoning index.

It is generally believed that January is the coldest month in Northern China, and the diurnal temperature range in January determines whether the concrete structure is damaged by freezing and thawing or not, while July is the hottest month, with the highest rate of carbonization and steel corrosion [22]. Therefore, the annual freeze-thaw cycles and the average temperature in July are selected as the zoning indexes.

Due to the different degree of industrialization, the annual emissions of acid gases such as SO_2, CO_2, and NO_x are different, resulting in different acid rainfall in each region. Therefore, the annual acid rainfall, known as the product of annual average precipitation and acid rain frequency [29], is selected as the zoning index.

3.3. Environmental Zonation for Concrete Durability

In the environmental zonation for concrete durability, the weight of the zoning index reflects the degree of its influence on concrete durability, and its rationality is directly related to the accuracy of zoning. In order to quantitatively zone the in-service environment of concrete structures in Shaanxi Province, this paper adopts the weight distribution method based on fuzzy and rough sets, which converts the weight determination problem into the attribute evaluation problem of rough sets, and the subjective influence of factor selection and weight determination can be avoided.

3.3.1. Fuzzy Cluster Analysis and the Theory of the Rough Set

Fuzzy Cluster Analysis

The traditional clustering analysis method is used to strictly divide each object into a certain category, but the attributes of objective things are usually not very clear and their categories are relatively vague, so it is effective to use the fuzzy clustering analysis to deal with the above-mentioned ambiguous objects. Fuzzy cluster analysis can obtain the degree of uncertainty of samples belonging to each category, and establish the uncertainty description, which can more objectively reflect the field situation [30].

For environmental zonation, let $\boldsymbol{X} = \{x_1, x_2, \ldots, x_n\}$ be the buildings to be tested, and these buildings will be divided into different categories according to their durability deterioration mechanism. Each sample $x_i = \{x_{i1}, x_{i2}, \ldots, x_{im}\}$ consists of m measured value of the durability influencing factors. The raw $n \times m$ data matrix can then be obtained.

The fuzzy clustering analysis is as follows [31,32]:

Step 1. Data standardization. In order to compare different testing values with different dimensions, the measured data should be standardized by *Standard Deviation Transformation* or *Range Transformation*.

$$x_{ij}' = \frac{x_{ij} - \mu_j}{s_j} \tag{1}$$

$$x_{ij}' = \frac{x_{ij} - \min\{x_{ij}\}}{\max\limits_{1\le i\le n}\{x_{ij}\} - \min\limits_{1\le i\le n}\{x_{ij}\}} \tag{2}$$

where μ_j and s_j is the mean value and the standard deviation of x_{ij}, respectively, and $i = 1, 2, \ldots, n$, $j = 1, 2, \ldots, m$.

Step 2. Establish the fuzzy similar matrix $\boldsymbol{R}$. The similar coefficient $r_{ij} = \boldsymbol{R}(x_i, x_j)$ between x_i and x_j can be calculated by using standardized data and the fuzzy similar matrix can be established.

Step 3. Establish the fuzzy equivalent matrix $\boldsymbol{R}^*$. Fuzzy similar matrix $\boldsymbol{R}$ is usually non-transitive, but reflexive and symmetrical. The equivalent matrix $\boldsymbol{R}^*$ can be calculated by using the transitive closure method. Stepwise calculate $\boldsymbol{R}^2, \boldsymbol{R}^4, \ldots$, until $\boldsymbol{R}^k = \boldsymbol{R}^{2k} = \boldsymbol{R}^2$.

Step 4. Clustering. Choose different confidence levels $\lambda \in [0, 1]$ and gradually merge the rows and columns of the fuzzy equivalent matrix $\boldsymbol{R}^*$ obtained in Step 3 to obtain different clustering results. When $r_{ij} \ge \lambda$, x_i and x_j are considered close enough to be the same category. For environmental zonation, it is considered that the influencing factors of concrete structure durability are similar. The larger the λ is, the higher the stability of sample elements is and the more details are distinguished. Therefore, the classification results are different when the confidence level is different. In view of this, the *F*-statistics method is used to determine the optimum threshold λ [31].

In the process of clustering analysis, the center vector can be obtained through a raw data matrix.

$$\overline{\boldsymbol{x}} = (\overline{x}_1, \overline{x}_2, \cdots, \overline{x}_k, \cdots, \overline{x}_m) \tag{3}$$

$$\overline{x}_k = \frac{1}{n}\sum_{i=1}^{n} x_{ik}, k = 1, 2, \cdots, m. \tag{4}$$

where $\overline{\boldsymbol{x}}$ is the center vector of the sample space.

Assume that the number of cluster is r when the confidence level is λ, and the j-th cluster has n_j samples, which are $x_1^{(j)}, x_2^{(j)}, \ldots, x_{n_j}^{(j)}$. Thus, the center vector of the j-th cluster is $\overline{\boldsymbol{x}}^{(j)} = (\overline{x}_1{}^{(j)}, \overline{x}_2{}^{(j)}, \cdots, \overline{x}_k{}^{(j)}, \cdots, \overline{x}_m{}^{(j)})$, where $\overline{x}_k{}^{(j)}$ is the average value of the k-th durability factor.

$$\overline{x}_k{}^{(j)} = \frac{1}{n_j}\sum_{i=1}^{n_j} x_{ik}{}^{(j)}, k = 1, 2, \cdots, m. \tag{5}$$

The value of *F*-statistic is

$$\mathbf{F} = \frac{\sum\limits_{j=1}^{r} n_j \|\overline{\boldsymbol{x}}^{(j)} - \overline{\boldsymbol{x}}\|^2 / (r-1)}{\sum\limits_{j=1}^{r}\sum\limits_{i=1}^{n_j} \|\boldsymbol{x}_i^{(j)} - \overline{\boldsymbol{x}}^{(j)}\|^2 / (n-r)} \tag{6}$$

where $\|\overline{\boldsymbol{x}}^{(j)} - \overline{\boldsymbol{x}}\| = \sqrt{\sum\limits_{k=1}^{m} (\overline{x}_k{}^{(j)} - \overline{x}_k)^2}$ is the distance between $\overline{\boldsymbol{x}}^{(j)}$ and $\overline{\boldsymbol{x}}$, and $\|\boldsymbol{x}_i^{(j)} - \overline{\boldsymbol{x}}^{(j)}\|$ is the distance between $\boldsymbol{x}_i{}^{(j)}$ and center vector $\overline{\boldsymbol{x}}^{(j)}$ in the j-th class. The molecular of the *F*-statistic represents the distance between categories, and the denominator represents the distance between samples in the same category. Therefore, the larger the value of the *F*-statistic becomes, the greater the distance between categories is, and the better the cluster result will be.

Rough Set theory

Rough set theory, proposed by Z. Pawlak, a Polish mathematician in 1982, is a theoretical method for studying the expression, learning, and induction of incomplete, uncertain knowledge and data. Its main idea is to make determination or zonation through knowledge reduction while maintaining the same distinguishing ability. Rough set theory has been widely used in data mining, pattern recognition, machine learning, and intelligent control because it can solve problems without prior knowledge [33,34].

The in-service environment of concrete structures usually has no detailed description, so it is impossible to obtain the weight of each index by statistical analysis. In view of this, rough set theory is introduced to determine the significance of environmental factors affecting the durability of concrete structures from rough description [35,36].

Definition 1. *The knowledge representation system is defined as follows:*

$$S = (\boldsymbol{U}, A, V, f) \tag{7}$$

where $\boldsymbol{U}$ is the set of samples; $A = C \cup D$ is the set of attributes, and C and D are the conditional attributes and decision attributes of the samples respectively; $V = \underset{a \in A}{\cup} V_a$ is the set of attribute values; f: $\boldsymbol{U} \times (C \cup \boldsymbol{D}) \rightarrow V$ is an information function that specifies the attribute values of each sample in $\boldsymbol{U}$. Each attribute subset determines a binary indiscernible relationship IND ($\boldsymbol{R}$):

$$IND(\boldsymbol{R}) = \{(x, y) \in \boldsymbol{U} \times \boldsymbol{U} | \forall a \in \boldsymbol{R}, f(x, a) = f(y, a)\} \tag{8}$$

Definition 2. *Given the knowledge representation system* $\boldsymbol{S} = (\boldsymbol{U}, A, V, f)$, *for each subset* $\boldsymbol{X} \subset \boldsymbol{U}$ *and indiscernible relation* $\boldsymbol{R} \subset A$, *the upper and lower approximation sets of* $\boldsymbol{X}$ *are defined as follows:*

$$\boldsymbol{R}^{-}(\boldsymbol{X}) = \cup\{\boldsymbol{Y} \subset \boldsymbol{U}/IND(\boldsymbol{R})/|\boldsymbol{Y} \cap \boldsymbol{X} \neq \varnothing\} \tag{9}$$

$$\boldsymbol{R}_{-}(\boldsymbol{X}) = \cup\{\boldsymbol{Y} \subset \boldsymbol{U}/IND(\boldsymbol{R})|\boldsymbol{Y} \subset \boldsymbol{X}\} \tag{10}$$

Definition 3. *The dependence between the two attribute sets* $\boldsymbol{C}$ *and* $\boldsymbol{D}$ *is defined as*

$$\gamma(\boldsymbol{C}, \boldsymbol{D}) = |POS_C(\boldsymbol{D})|/|\boldsymbol{U}| \tag{11}$$

where $POS_C(\boldsymbol{D}) = C\_(\boldsymbol{D})$, and $|\boldsymbol{U}|$ is the number of the element of $\boldsymbol{U}$.

Definition 4. *Attribute* $a \in C$, *the significance of attribute* a *for* $\boldsymbol{D}$ *is defined as*

$$SGF(a, \boldsymbol{C}, \boldsymbol{D}) = \gamma(\boldsymbol{C}, \boldsymbol{D}) - \gamma(\boldsymbol{C} - \{a\}, \boldsymbol{D}) \tag{12}$$

where $\gamma(\boldsymbol{C} - \{a\}, \boldsymbol{D})$ represents the dependence of conditional attributes on the decision attributes after removing attribute a from $\boldsymbol{C}$.

3.3.2. Weight Distribution of the Index of Environmental Zonation for Concrete Durability

Before doing environmental zonation, the weight distribution of various indexes should be determined. Using rough set theory to determine the importance of each zoning index is to calculate the weight of each index by attribute reduction under the premise of maintaining the classification ability. Firstly, the zoning index is fuzzy clustered, and the best classification can be obtained. The index c_i is then successively deleted from the classification and the remaining indexes are clustered with the same method. The significance of the index was defined by the change of the whole system after deleting the index. That is, the more significant the index is, the more important it is to maintain the stability of the whole system, and the greater its weight becomes. The specific steps are as follows:

Step 1. Take all test samples as the universe $\boldsymbol{U} = \{x_1, x_2, \ldots, x_n\}$, the index of environmental zonation $(c_{i1}, c_{i2}, \ldots, c_{im})$ as conditional attributes, and the durability grade of structure or component $(d_1, d_2, \ldots, d_m)$ as the decision attribute. The raw data matrix can then be obtained:

$$\boldsymbol{X} = \begin{bmatrix} c_{11} & c_{12} & \cdots & c_{1m} & d_1 \\ c_{21} & c_{22} & \cdots & c_{2m} & d_2 \\ \vdots & \vdots & \ddots & \vdots & \vdots \\ c_{n1} & c_{n2} & \cdots & c_{nm} & d_m \end{bmatrix} \tag{13}$$

Step 2. Determine the optimal confidence level λ by the F-statistics method, and then divide the equivalent classification of the tested buildings according to the zoning index and durability damage level, respectively, to find the best classification:

$$\boldsymbol{Y} = \{\boldsymbol{Y}_1, \boldsymbol{Y}_2, \ldots, \boldsymbol{Y}_s\} \tag{14}$$

where $\boldsymbol{Y}_i$ is an equivalent set, which is a set of equivalent relations corresponding to a certain durability damage level.

Step 3. Delete the index c_i $(i = 1, 2, \ldots, m)$ and use the aforementioned method to process the remaining matrix, and a new classification set after removing the index c_i can then be achieved.

$$\boldsymbol{E} = \{\boldsymbol{E}_1, \boldsymbol{E}_2, \ldots, \boldsymbol{E}_m\} \tag{15}$$

where $\boldsymbol{E}_i = \{\boldsymbol{Y}_1^{(i)}, \boldsymbol{Y}_2^{(i)}, \ldots, \boldsymbol{Y}_k^{(i)}\}$ is the equivalent set obtained after removing the i-th zoning index. For different i, k may be different. $\boldsymbol{Y}_l^{(i)}$ $(1 \le l \le k)$ is the l-th equivalent set obtained after removing the i-th zoning index.

Step 4. The importance of each index is obtained by rough set theory. According to Definition 2, the union of the lower approximation set of each equivalent set for each durability damage level is obtained.

$$POS_{C-\{c_i\}}(\boldsymbol{D}) = \{\boldsymbol{C} - \{c_i\}\}\_(\boldsymbol{D}) = \cup\{\{\boldsymbol{C} - \{c_i\}\}\_\boldsymbol{Y}_l\} \tag{16}$$

For each zoning index c_i, the dependence of durability damage level $\boldsymbol{D}$ on zoning index set $\boldsymbol{C}$ and zoning index set $\boldsymbol{C} - \{c_i\}$ can be calculated by Definition 3 of rough set theory.

$$\gamma(\boldsymbol{C}, \boldsymbol{D}) = |POS_C(\boldsymbol{D})|/|\boldsymbol{U}| \tag{17}$$

$$\gamma(\boldsymbol{C} - \{c_i\}, \boldsymbol{D}) = |POS_{C-\{c_i\}}(\boldsymbol{D})|/|\boldsymbol{U}| \tag{18}$$

The significance $\boldsymbol{SGF}(c_i, \boldsymbol{C}, \boldsymbol{D})$ of zoning index c_i can then be calculated by Definition 4 of rough set theory.

Step 5. According to the significance of each zoning index, the weight of each index is normalized to get more intuitive results.

$$\boldsymbol{W}_i = SGF(c_i, \boldsymbol{C}, \boldsymbol{D}) / \sum_{k=1}^{m} SGF(c_k, \boldsymbol{C}, \boldsymbol{D}) \tag{19}$$

To summarize, the knowledge system is established through the establishment of a relational data model and the characterization of an attribute value, and the factor weight is then calculated by analyzing the support degree and significance of the evaluation object under the data driven. This is a weight assignment method of the index of environmental zonation proposed in this study.

3.3.3. Environmental Zonation for Concrete Durability in Shaanxi Province

A large number of buildings have been built in Shaanxi Province in recent years. However, due to the insufficient understanding of the durability of concrete structures in the early years, durability problems of concrete structures are quite common. This study tested the durability influencing factors of 989 existing concrete structures in Shaanxi Province and assessed the durability damage levels of them by taking into account different construction years, structural types, and geographical locations. Take Yulin as an example to illustrate the weight calculation. Twenty test samples are selected as the universe $\boldsymbol{U} = \{1, 2, \ldots, 20\}$. The conditional attribute set $\boldsymbol{C} = \{c_1, c_2, c_3, c_4\}$ includes annual freeze-thaw cycles, annual precipitation, the average temperature in July, and the annual acid rainfall. The decision attribute set $\boldsymbol{D} = \{D_1\}$ is the durability damage level of sample components, which can be calculated according to the *Standard for Durability Assessment of Existing Concrete Structures* (GB/T 51355-2019) [37]. In the decision attribute set, Durability Damage Level 1 indicates that the component has no durability damage, Level 2 indicates that the component has slight mechanical damage or durability damage, Level 3 indicates that the component has more serious durability damage, and Level 4 indicates that the component has very serious durability damage. The test data are shown in Table 2.

Table 2. Detecting data of concrete structures (extract).

Sample Number	Indexes of Environmental Zonation for Concrete Durability				
	Annual Freeze-Thaw Cycles	Annual Precipitation (mm)	Average Temperature in July (°C)	Annual Acid Rainfall (mm)	Durability Damage Level
1	113	574.9	23.9	91.9	4
2	110	562.2	23.5	101.1	3
3	106	538.9	23.3	106.1	3
4	117	567.1	23.5	97.4	4
5	114	542.2	23.8	103.7	3
6	114	547.6	23.5	105.1	4
7	103	515.2	23.1	113.7	2
8	113	572.0	23.5	95.5	4
9	108	544.7	22.8	107.2	3
10	109	546.1	24.2	102.0	3
11	112	570.2	23.4	94.3	4
12	106	521.3	23.3	107.6	2
13	107	505.5	23.7	112.0	1
14	105	517.7	23.4	110.6	2
15	115	578.5	23.6	95.3	4
16	104	505.1	23.0	111.3	1
17	116	543.5	23.5	101.9	4
18	108	566.8	23.2	100.8	3
19	108	510.3	23.6	111.3	2
20	115	545.6	23.3	106.7	3
$\bar{x}_k$	110	543.8	23.4	103.8	3
$\bar{s}_k$	4.37	23.69	0.30	6.33	0.97

(1) Establish fuzzy similarity matrix

The detecting data are then standardized by standard deviation to eliminate the influence of dimension, and the standardization matrix $\boldsymbol{R}$ is obtained.

$$\begin{bmatrix} 0.74 & 1.31 & 1.33 & -1.87 & 1.08 \\ 0.06 & 0.78 & 0.02 & -0.43 & 0.05 \\ -1.06 & -0.21 & -0.57 & 0.37 & 0.05 \\ 1.66 & 0.99 & 0.31 & -1.01 & 1.08 \\ 0.77 & -0.07 & 1.04 & -0.01 & 0.05 \\ 0.79 & 0.16 & 0.12 & 0.21 & 1.08 \\ -1.71 & -1.20 & -1.07 & 1.57 & -0.98 \\ 0.76 & 1.19 & 0.28 & -1.30 & 1.08 \\ -0.57 & 0.04 & -2.25 & 0.55 & 0.05 \\ -0.17 & 0.10 & 2.50 & -0.28 & 0.05 \\ 0.53 & 1.11 & -0.04 & -1.49 & 1.08 \\ -0.85 & -0.95 & -0.39 & 0.61 & -0.98 \\ -0.83 & -1.62 & 0.73 & 1.30 & -2.00 \\ -1.26 & -1.10 & -0.22 & 1.08 & -0.98 \\ 1.02 & 1.47 & 0.34 & -1.33 & 1.08 \\ -1.52 & -1.63 & -1.36 & 1.18 & -2.00 \\ 1.41 & -0.01 & 0.16 & -0.30 & 1.08 \\ -0.39 & 0.97 & -0.92 & -0.48 & 0.05 \\ -0.58 & -1.41 & 0.52 & 1.19 & -0.98 \\ 1.20 & 0.08 & -0.55 & 0.46 & 0.05 \end{bmatrix}.$$

Use the Euler distance method to establish similarity matrix $\boldsymbol{R}$,

$$r_{ij} = 1 - C \cdot \sqrt{\sum_{k=1}^{5} (x_{ik} - x_{jk})^2}$$

To ensure $0 \le r_{ij} \le 1$, let $C = 0.125$ and the fuzzy similarity matrix $\boldsymbol{R}$ can be obtained.

$$\left[\begin{array}{cccccccccccccccccccc}
1 & 0.734 & 0.529 & 0.793 & 0.708 & 0.666 & 0.316 & 0.850 & 0.413 & 0.689 & 0.818 & 0.488 & 0.421 & 0.428 & 0.853 & 0.312 & 0.693 & 0.638 & 0.453 & 0.591 \\
0.734 & 1 & 0.776 & 0.783 & 0.805 & 0.855 & 0.563 & 0.848 & 0.668 & 0.676 & 0.849 & 0.718 & 0.604 & 0.655 & 0.810 & 0.553 & 0.803 & 0.867 & 0.644 & 0.787 \\
0.529 & 0.776 & 1 & 0.576 & 0.691 & 0.748 & 0.780 & 0.630 & 0.778 & 0.590 & 0.646 & 0.897 & 0.732 & 0.849 & 0.588 & 0.766 & 0.667 & 0.795 & 0.765 & 0.715 \\
0.793 & 0.783 & 0.576 & 1 & 0.768 & 0.785 & 0.379 & 0.879 & 0.518 & 0.615 & 0.840 & 0.546 & 0.462 & 0.477 & 0.892 & 0.381 & 0.842 & 0.694 & 0.505 & 0.752 \\
0.708 & 0.805 & 0.691 & 0.768 & 1 & 0.878 & 0.526 & 0.755 & 0.550 & 0.779 & 0.726 & 0.698 & 0.675 & 0.647 & 0.730 & 0.518 & 0.859 & 0.681 & 0.711 & 0.785 \\
0.666 & 0.855 & 0.748 & 0.785 & 0.878 & 1 & 0.579 & 0.770 & 0.655 & 0.673 & 0.753 & 0.740 & 0.661 & 0.678 & 0.744 & 0.573 & 0.897 & 0.763 & 0.708 & 0.896 \\
0.316 & 0.563 & 0.780 & 0.379 & 0.526 & 0.579 & 1 & 0.415 & 0.713 & 0.437 & 0.430 & 0.815 & 0.742 & 0.864 & 0.375 & 0.915 & 0.498 & 0.591 & 0.750 & 0.574 \\
0.850 & 0.848 & 0.630 & 0.879 & 0.755 & 0.770 & 0.415 & 1 & 0.551 & 0.646 & 0.944 & 0.580 & 0.479 & 0.512 & 0.952 & 0.414 & 0.788 & 0.767 & 0.519 & 0.714 \\
0.413 & 0.668 & 0.778 & 0.518 & 0.550 & 0.655 & 0.713 & 0.551 & 1 & 0.395 & 0.578 & 0.734 & 0.562 & 0.689 & 0.519 & 0.724 & 0.596 & 0.759 & 0.601 & 0.693 \\
0.689 & 0.676 & 0.590 & 0.615 & 0.779 & 0.673 & 0.437 & 0.646 & 0.395 & 1 & 0.616 & 0.591 & 0.624 & 0.569 & 0.624 & 0.415 & 0.647 & 0.557 & 0.635 & 0.572 \\
0.818 & 0.849 & 0.646 & 0.840 & 0.726 & 0.753 & 0.430 & 0.944 & 0.578 & 0.616 & 1 & 0.591 & 0.474 & 0.520 & 0.908 & 0.433 & 0.767 & 0.796 & 0.514 & 0.705 \\
0.488 & 0.718 & 0.897 & 0.546 & 0.698 & 0.740 & 0.815 & 0.580 & 0.734 & 0.591 & 0.591 & 1 & 0.815 & 0.917 & 0.538 & 0.815 & 0.667 & 0.710 & 0.849 & 0.712 \\
0.421 & 0.604 & 0.732 & 0.462 & 0.675 & 0.661 & 0.742 & 0.479 & 0.562 & 0.624 & 0.474 & 0.815 & 1 & 0.852 & 0.440 & 0.724 & 0.595 & 0.553 & 0.950 & 0.618 \\
0.428 & 0.655 & 0.849 & 0.477 & 0.647 & 0.678 & 0.864 & 0.512 & 0.689 & 0.569 & 0.520 & 0.917 & 0.852 & 1 & 0.471 & 0.839 & 0.598 & 0.647 & 0.868 & 0.648 \\
0.853 & 0.810 & 0.588 & 0.892 & 0.730 & 0.744 & 0.375 & 0.952 & 0.519 & 0.624 & 0.908 & 0.538 & 0.440 & 0.471 & 1 & 0.372 & 0.768 & 0.733 & 0.481 & 0.695 \\
0.312 & 0.553 & 0.766 & 0.381 & 0.518 & 0.573 & 0.915 & 0.414 & 0.724 & 0.415 & 0.433 & 0.815 & 0.724 & 0.839 & 0.372 & 1 & 0.505 & 0.585 & 0.736 & 0.576 \\
0.693 & 0.803 & 0.667 & 0.842 & 0.859 & 0.897 & 0.498 & 0.788 & 0.596 & 0.647 & 0.767 & 0.667 & 0.595 & 0.598 & 0.768 & 0.505 & 1 & 0.710 & 0.641 & 0.867 \\
0.638 & 0.867 & 0.795 & 0.694 & 0.681 & 0.763 & 0.591 & 0.767 & 0.759 & 0.557 & 0.796 & 0.710 & 0.553 & 0.647 & 0.733 & 0.585 & 0.710 & 1 & 0.593 & 0.740 \\
0.453 & 0.644 & 0.765 & 0.505 & 0.711 & 0.708 & 0.750 & 0.519 & 0.601 & 0.635 & 0.514 & 0.849 & 0.950 & 0.868 & 0.481 & 0.736 & 0.641 & 0.593 & 1 & 0.668 \\
0.591 & 0.787 & 0.715 & 0.752 & 0.785 & 0.896 & 0.574 & 0.714 & 0.693 & 0.572 & 0.705 & 0.712 & 0.618 & 0.648 & 0.695 & 0.576 & 0.867 & 0.740 & 0.668 & 1
\end{array}\right]$$

(2) Clustering

Use the square method to find the transitive closure, and the fuzzy equivalent matrix R^* can be obtained:

$$\begin{bmatrix}
1 & 0.849 & 0.795 & 0.853 & 0.849 & 0.849 & 0.795 & 0.853 & 0.778 & 0.779 & 0.853 & 0.795 & 0.795 & 0.795 & 0.853 & 0.795 & 0.849 & 0.849 & 0.795 & 0.849 \\
0.849 & 1 & 0.795 & 0.849 & 0.855 & 0.855 & 0.795 & 0.849 & 0.778 & 0.779 & 0.849 & 0.795 & 0.795 & 0.795 & 0.849 & 0.795 & 0.855 & 0.867 & 0.795 & 0.855 \\
0.795 & 0.795 & 1 & 0.795 & 0.795 & 0.795 & 0.864 & 0.795 & 0.778 & 0.779 & 0.795 & 0.897 & 0.868 & 0.897 & 0.795 & 0.864 & 0.795 & 0.795 & 0.868 & 0.795 \\
0.853 & 0.849 & 0.795 & 1 & 0.849 & 0.849 & 0.795 & 0.892 & 0.778 & 0.779 & 0.892 & 0.795 & 0.795 & 0.795 & 0.892 & 0.795 & 0.849 & 0.849 & 0.795 & 0.849 \\
0.849 & 0.855 & 0.795 & 0.849 & 1 & 0.878 & 0.795 & 0.849 & 0.778 & 0.779 & 0.849 & 0.795 & 0.795 & 0.795 & 0.849 & 0.795 & 0.878 & 0.855 & 0.795 & 0.878 \\
0.849 & 0.855 & 0.795 & 0.849 & 0.878 & 1 & 0.795 & 0.849 & 0.778 & 0.779 & 0.849 & 0.795 & 0.795 & 0.795 & 0.849 & 0.795 & 0.897 & 0.855 & 0.795 & 0.896 \\
0.795 & 0.795 & 0.864 & 0.795 & 0.795 & 0.795 & 1 & 0.795 & 0.778 & 0.779 & 0.795 & 0.864 & 0.864 & 0.864 & 0.795 & 0.915 & 0.795 & 0.795 & 0.864 & 0.795 \\
0.853 & 0.849 & 0.795 & 0.892 & 0.849 & 0.849 & 0.795 & 1 & 0.778 & 0.779 & 0.944 & 0.795 & 0.795 & 0.795 & 0.952 & 0.795 & 0.849 & 0.849 & 0.795 & 0.849 \\
0.778 & 0.778 & 0.778 & 0.778 & 0.778 & 0.778 & 0.778 & 0.778 & 1 & 0.778 & 0.778 & 0.778 & 0.778 & 0.778 & 0.778 & 0.778 & 0.778 & 0.778 & 0.778 & 0.778 \\
0.779 & 0.779 & 0.779 & 0.779 & 0.779 & 0.779 & 0.779 & 0.779 & 0.778 & 1 & 0.779 & 0.779 & 0.779 & 0.779 & 0.779 & 0.779 & 0.779 & 0.779 & 0.779 & 0.779 \\
0.853 & 0.849 & 0.795 & 0.892 & 0.849 & 0.849 & 0.795 & 0.944 & 0.778 & 0.779 & 1 & 0.795 & 0.795 & 0.795 & 0.944 & 0.795 & 0.849 & 0.849 & 0.795 & 0.849 \\
0.795 & 0.795 & 0.897 & 0.795 & 0.795 & 0.795 & 0.864 & 0.795 & 0.778 & 0.779 & 0.795 & 1 & 0.868 & 0.917 & 0.795 & 0.864 & 0.795 & 0.795 & 0.868 & 0.795 \\
0.795 & 0.795 & 0.868 & 0.795 & 0.795 & 0.795 & 0.864 & 0.795 & 0.778 & 0.779 & 0.795 & 0.868 & 1 & 0.868 & 0.795 & 0.864 & 0.795 & 0.795 & 0.95 & 0.795 \\
0.795 & 0.795 & 0.897 & 0.795 & 0.795 & 0.795 & 0.864 & 0.795 & 0.778 & 0.779 & 0.795 & 0.917 & 0.868 & 1 & 0.795 & 0.864 & 0.795 & 0.795 & 0.868 & 0.795 \\
0.853 & 0.849 & 0.795 & 0.892 & 0.849 & 0.849 & 0.795 & 0.952 & 0.778 & 0.779 & 0.944 & 0.795 & 0.795 & 0.795 & 1 & 0.795 & 0.849 & 0.849 & 0.795 & 0.849 \\
0.795 & 0.795 & 0.864 & 0.795 & 0.795 & 0.795 & 0.915 & 0.795 & 0.778 & 0.779 & 0.795 & 0.864 & 0.864 & 0.864 & 0.795 & 1 & 0.795 & 0.795 & 0.864 & 0.795 \\
0.849 & 0.855 & 0.795 & 0.849 & 0.878 & 0.897 & 0.795 & 0.849 & 0.778 & 0.779 & 0.849 & 0.795 & 0.795 & 0.795 & 0.849 & 0.795 & 1 & 0.855 & 0.795 & 0.896 \\
0.849 & 0.867 & 0.795 & 0.849 & 0.855 & 0.855 & 0.795 & 0.849 & 0.778 & 0.779 & 0.849 & 0.795 & 0.795 & 0.795 & 0.849 & 0.795 & 0.855 & 1 & 0.795 & 0.855 \\
0.795 & 0.795 & 0.868 & 0.795 & 0.795 & 0.795 & 0.864 & 0.795 & 0.778 & 0.779 & 0.795 & 0.868 & 0.95 & 0.868 & 0.795 & 0.864 & 0.795 & 0.795 & 1 & 0.795 \\
0.849 & 0.855 & 0.795 & 0.849 & 0.878 & 0.896 & 0.795 & 0.849 & 0.778 & 0.779 & 0.849 & 0.795 & 0.795 & 0.795 & 0.849 & 0.795 & 0.896 & 0.855 & 0.795 & 1
\end{bmatrix}$$

Use the F-statistics method to determine the optimal threshold value of $\lambda = 0.915$ and classify the samples. The classification results of the conditional attribute and the decision attribute are as follows:

$U/IND(\boldsymbol{C})$= {{1},{2},{3},{4},{5},{6},{7,16},{8,11,15},{9},{10},{12,14},{13,19},{17},{18},{20}};

$U/IND(\boldsymbol{D})$= {{13,16},{7,12,14,19},{2,3,5,9,10,18,20},{1,4,6,8,11,15,17}}.

Similarly, after removing the c_i conditional attributes, the fuzzy clustering method is used to classify the remaining conditional attributes, and the best classifications are as follows:

$U/IND(\boldsymbol{C}-\{c_1\})$ = {{1},{2,4,8,11,15},{3,20},{5},{6,17},{7,16},{9},{10},{12,14},{13,19},{18}};

$U/IND(\boldsymbol{C}-\{c_2\})$ = {{1},{2},{3,12,14},{4,8,11,15},{5},{6},{7,16},{9},{10},{13,19},{17},{18},{20}};

$U/IND(\boldsymbol{C}-\{c_3\})$ = {{1,8,11,15},{2,18},{3},{4},{5,6,20},{7},{9},{10},{12},{13,19},{14},{16},{17}}.

$U/IND(\boldsymbol{C}-\{c_4\})$ = {{1},{2,8,11,15},{3},{4},{5},{6},{7,16},{9},{10},{12,14},{13,19},{17},{18},{20}}.

Calculate the importance of each conditional attribute to the decision attribute:

$$POS_C(\boldsymbol{D}) = \{1,2,3,4,5,6,8,9,10,11,12,14,15,17,18,20\}$$

$$POS_{C-\{c_1\}}(\boldsymbol{D}) = \{1,3,5,6,17,9,10,12,14,18,20\}$$

$$POS_{C-\{c_2\}}(\boldsymbol{D}) = \{1,2,4,5,6,8,9,10,11,15,17,18,20\}$$

$$POS_{C-\{c_3\}}(\boldsymbol{D}) = \{1,2,3,4,7,8,9,10,11,12,14,15,16,17,18\}$$

$$POS_{C-\{c_4\}}(\boldsymbol{D}) = \{1,3,4,5,6,9,10,12,14,17,18,20\}$$

Calculate the significance $SGF(c_i, \boldsymbol{C}, \boldsymbol{D})$ of each zoning index c_i through the dependence degree of durability damage level $\boldsymbol{D}$ on the zoning index set $\boldsymbol{C}$ and the zoning index set $\boldsymbol{C}-c_i$.

$$SGF(c_1, \boldsymbol{C}, \boldsymbol{D}) = \gamma_C(\boldsymbol{D}) - \gamma_{C-\{c_1\}}(\boldsymbol{D}) = \frac{|POS_C(\boldsymbol{D})|}{|\boldsymbol{U}|} - \frac{|POS_{C-\{c_1\}}(\boldsymbol{D})|}{|\boldsymbol{U}|} = \frac{16}{20} - \frac{11}{20} = \frac{5}{20}$$

$$SGF(c_2, \boldsymbol{C}, \boldsymbol{D}) = \gamma_C(\boldsymbol{D}) - \gamma_{C-\{c_2\}}(\boldsymbol{D}) = \frac{|POS_C(\boldsymbol{D})|}{|\boldsymbol{U}|} - \frac{|POS_{C-\{c_2\}}(\boldsymbol{D})|}{|\boldsymbol{U}|} = \frac{16}{20} - \frac{13}{20} = \frac{3}{20}$$

$$SGF(c_3, \boldsymbol{C}, \boldsymbol{D}) = \gamma_C(\boldsymbol{D}) - \gamma_{C-\{c_3\}}(\boldsymbol{D}) = \frac{|POS_C(\boldsymbol{D})|}{|\boldsymbol{U}|} - \frac{|POS_{C-\{c_3\}}(\boldsymbol{D})|}{|\boldsymbol{U}|} = \frac{16}{20} - \frac{15}{20} = \frac{1}{20}$$

$$SGF(c_4, \boldsymbol{C}, \boldsymbol{D}) = \gamma_C(\boldsymbol{D}) - \gamma_{C-\{c_4\}}(\boldsymbol{D}) = \frac{|POS_C(\boldsymbol{D})|}{|\boldsymbol{U}|} - \frac{|POS_{C-\{c_4\}}(\boldsymbol{D})|}{|\boldsymbol{U}|} = \frac{16}{20} - \frac{12}{20} = \frac{4}{20}$$

Finally, the weight of each zoning index is obtained after normalizing the significance.

$$(\omega_{c_1}, \omega_{c_2}, \omega_{c_3}, \omega_{c_4}) = (0.3846, 0.2308, 0.0769, 0.3077) \tag{20}$$

It can be seen that, among the factors affecting the durability of concrete structures in Yulin, the annual freeze-thaw cycles has the greatest impact on the durability of concrete, followed by annual acid rain and annual precipitation. The average temperature in July has a relatively small impact. Based on the above methods, the factors affecting the durability of 989 existing concrete buildings in typical cities of Shaanxi Province were analyzed, and their durability statuses were assessed. The weight of each factor and the durability damage level of the buildings were obtained. Considering the practicability of the zoning results, the environmental zonation was carried out by taking into account the landform, the road network, the administrative division, and the durability deterioration mechanism of each zone in Shaanxi Province. Based on the principle of *a smaller difference within the region, a larger difference between regions* and *regional conjugation* [38], Shaanxi Province was divided into three first-level zones and subdivided into nine second-level zones by the weight of zoning indexes. Detailed zoning result is shown in Table 3. According to the zoning result, the durability zoning map of concrete structures in Shaanxi Province can be plotted as Figure 9.

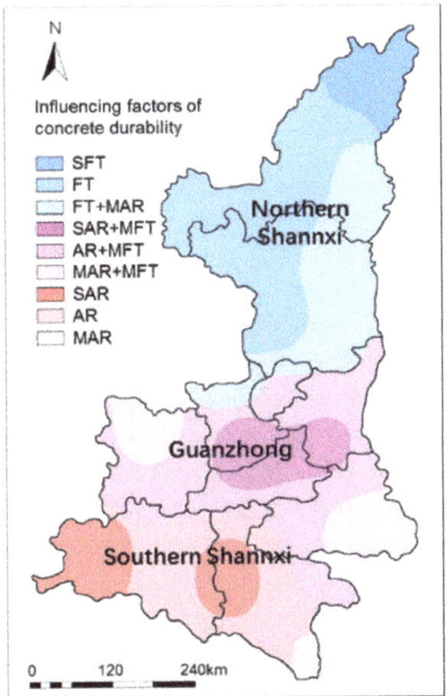

Figure 9. Durability zoning map of concrete structures in Shaanxi Province, SFT, FT, MFT, SAR, AR, MAR = severe freeze-thaw, moderate freeze-thaw, mild freeze-thaw, severe acid rain, moderate acid rain, and mild acid rain.

Table 3. Result of environmental zonation for concrete durability in Shaanxi Province.

Zones		Weight of the Indexes for Environmental Zonation				Typical City
		Annual Freeze-Thaw Cycles	Annual Precipitation	Average Temperature in July	Annual Acid Rainfall	
Freeze-Thaw Cycle Zone (Zone I)	Severe freeze-thaw (Zone I_a) [1]	0.4231~0.5170	0.1722~0.2358	0.1152~0.1875	0.0202~0.2003	Fugu, Shenmu
	Moderate freeze-thaw (Zone I_b)	0.3076~0.4611	0.1361~0.2404	0.0221~0.1813	0.1018~0.2384	Yulin, Hengshan, Jingbian, Dingbian, Wuqi, Zhidan, Ansai, Ganquan, Fuxian
	Moderate freeze-thaw with mild acid rain (Zone I_c)	0.2845~0.4212	0.1533~0.2398	0.1421~0.1943	0.1592~0.2822	Suide, Mizhi, Jiaxian, Wubao, Qingjian, Zizhou, Zichang, Yanchuan, Yanchang, Yan'an, Yichuan, Luochuan, Huanglong, Huangling, Changwu, Xunyi, Yijun, Binxian.
Neutralization–Freeze-Thaw Interaction Zone (Zone II)	Severe acid rain with mild freeze-thaw (Zone II_a)	0.1462~0.2572	0.2082~0.2810	0.0576~0.1652	0.3512~0.4675	Xi'an, Xianyang, Xingping, Tongchuan, Weinan, Gaoling, Zhouzhi, Huxian, Lantian, Sanyuan, Jingyang, Qianxian, Liquan, Huaxian, Fuping.
	Moderate acid rain with mild freeze-thaw (Zone II_b)	0.2195~0.3023	0.2124~0.2708	0.1199~0.2124	0.2764~0.3743	Baoji, Taibai, Wugong, Qishan, Fufeng, Meixian, Yongshou, Chunhua, Hancheng, Heyang, Dali, Pucheng, Baishui, Chengcheng, Huayin, Tongguan, Xunyang, Baihe, Shangluo, Zhen'an, Zhashui, Luonan, Danfeng, Shanyang.
	Mild acid rain with mild freeze-thaw (Zone II_c)	0.2023~0.2863	0.2129~0.2545	0.1741~0.2318	0.2036~0.3201	Longxian, Qianyang, Fengxiang, Linyou, Fengxian, Shangnan.
Neutralization Zone (Zone III)	Severe acid rain (Zone III_a)	0.1097~0.1948	0.1245~0.2941	0.0785~0.1088	0.4425~0.5249	Ningqiang, Lueyang, Mianxian, Hanyin, Shiquan, Ningshan.
	Moderate acid rain (Zone III_b)	0.1362~0.2316	0.2198~0.3753	0.1856~0.2851	0.3518~0.4699	Hanzhong, Liuba, Nanzheng, Chenggu, Yangxian, Xixiang, Zhenba, Foping, Ankang, Ziyang, Langao, Pingli
	Mild acid rain (Zone III_c)	0.1037~0.1232	0.3495~0.3772	0.2433~0.2678	0.3247~0.4012	Zhenping

[1] Each first-level zone is subdivided into three second-level zones, namely I_a, I_b, I_c, etc.

3.3.4. Environmental Conditions of Each Zone

Table 3 and Figure 9 show that, from the Freeze-Thaw Cycle Zone (Zone I) to the Neutralization Zone (Zone III), the influence of the freeze-thaw cycle on the durability of existing concrete structures is gradually weakened, while the influence of neutralization gradually becomes strong. Based on this, the environmental factors and their effects on the durability of concrete structures in different zones are comprehensively analyzed.

Freeze-Thaw Cycle Zone (Zone I)

Zone I includes Yulin, Yan'an, Northern Xianyang, and Northern Tongchuan. It has a long and cold winter with an annual average temperature of 6.5–8.7 °C, annual precipitation of 330–700 mm, and annual sunshine duration of 1550–2820 h. In this zone, the average temperature in January is −10.5—5.0 °C, and the minimum temperature is −20–−30 °C, while the average temperature in July is 17.0–23.8 °C, and the maximum temperature is 35–44 °C. The annual frost-free period is 140–160 d, and the relative humidity in June–August is 58–76%. The annual range of air temperature can reach 26–34 °C, and the annual average daily range of air temperature is 7–14 °C. Acid rain occasionally falls in the southeast of this zone, with a frequency of less than 10%, and the maximum frozen soil depth is less than 1.2 m.

Concrete structures located in Zone I suffer from freeze-thaw cycles in winter and carbonization all year. The coupling effect of freeze-thaw and carbonization should be considered in the durability design of concrete structures. Moreover, in the southeast of this zone, the impact of occasional acid rain should be taken into account in the durability design of the building's external components.

Neutralization–Freeze-Thaw Interaction Zone (Zone II)

Zone II includes Baoji, Xi'an, Weinan, most of Shangluo, south of Xianyang and Tongchuan, and northeast of Ankang. The average annual temperature in this zone is 8.8~14.0 °C, and the average temperature in January and July are −7.7~−0.8 °C and 14.8~27.5 °C, respectively. This zone enjoys 154~235 annual frost-free days and 1300~2600 h annual sunshine duration. The annual precipitation in this zone is 450~1400 mm, and the relative humidity in June-August is 60~83%.

Concrete structures located in this zone are mainly subjected to neutralization and occasional freeze-thaw damage. In Zone II_c, such as Fengxiang, the frequency of acid rain is relatively low, so in this zone it is not suitable to take acid rain as a key factor in durability design. The acid rain frequency in Xi'an, Weinan, Tongchuan, and other places located in Zone II_a is more than 30%, or even up to 50~80% in Xi'an and Xianyang. The erosion of acid rain should be considered as a key factor in the durability design of concrete structures in this zone.

Neutralization Zone (Zone III)

Zone III mainly includes Hanzhong and most of Ankang. The average annual temperature in this zone is 14.2~16.3 °C, with a small daily range of temperature. The annual precipitation is 800~1100 mm, and the annual sunshine duration is 1550~1700 h. In most of the zone, it is sultry in summer but wet and cold in winter, with an average temperature of 25.0~27.5 °C in July and −0.7~−0.6 °C in January, respectively. The annual frost-free period is 231~253 d, and the relative humidity from June to August is 74~80%. The annual number of rainfall days is about 150 d, and some areas can exceed 200 d, and the annual snowfall days is 1~14 d.

In Zone III, the frequency of acid rain is over 30%. The durability of concrete structures in this zone is less affected by freeze-thaw cycles. The durability damage of concrete structures is mainly caused by concrete neutralization and becomes serious when the concrete components suffer from acid rain erosion. Zone III_a, represented by Ningqiang and Lueyang, is a typical acid rain disaster zone whose acid rain frequency is up to 80%. Serious acid rain can accelerate the neutralization of concrete, destroy the passive film on the surface of reinforcing bars, make the reinforcing bars rust prematurely, and cause structural damage.

4. Discussion

The environmental zonation for concrete durability is a complex problem of multi-index evaluation [39]. In order to quantitatively solve this problem, the fuzzy clustering analysis and rough set theory are involved, and the quantitative calculation method of the weight of the zoning index is proposed. Compared with the existing research results [16–18], the method proposed in this paper puts forward clear zoning indexes and quantifies them, which can avoid the subjectivity of the selection and evaluation of the zoning indexes, and improve the quality and accuracy of the zoning results. Meanwhile, the method proposed can scientifically zone the building environment of a specific region, and can comprehensively reflect the influence degree of durability factors.

Based on the method established in this paper, this study takes Shaanxi Province as an example and zones its environment, and the zoning results are well consistent with the durability deterioration of concrete structures, which proves the rationality, effectiveness, and practicability of this method. Thus, it can provide a scientific basis for the durability design and maintenance of reinforced concrete structures.

However, due to the complexity of the environment affecting the durability of concrete structures, the meteorological data and durability data collected are not sufficient, and the soil environment, water environment, and special corrosive environment are not analyzed. For a specific situation, a specific analysis is needed. For instance, the steel corrosion caused by chloride ion erosion should be considered in a coastal or deicing salt environment, and the influence of an alkali-aggregate reaction, delayed ettringite formation, and soft water erosion on durability should be considered when the concrete structures are chronically wet or exposed to water.

With the accumulation of data and the change of environment, zonation may change dynamically and become more precise. The follow-up research can establish the durability database to collect, arrange, systematize, and standardize the data, which is conducive to the durability design, evaluation, and life prediction of concrete structures.

5. Conclusions

Based on the investigation and analysis of the environmental conditions and the durability status of concrete structures in Shaanxi Province, this paper puts forward the zoning indexes and carries out the environmental zonation for concrete durability. The main conclusions are as follows:

(1) The environmental conditions of concrete structures in Shaanxi Province were investigated, including atmospheric temperature, precipitation, corrosive gases in the atmosphere, and acid rain. Generally speaking, the annual average temperature in Shaanxi Province is higher in the south and lower in the north, which has increased slightly in the past half century. The annual precipitation is more in the south than that in the north, and has decreased slightly in recent years. The corrosive gas emissions show an upward trend before 2011 and a downward trend after 2011, which is closely related to economic development and environmental governance.

(2) According to the survey results of durability of existing industrial buildings, residential buildings, roads, and bridges in Shaanxi Province, the main factors affecting the durability of concrete structures are freeze-thaw cycles and the concrete neutralization caused by carbonation and acid rain erosion. According to the deterioration mechanism of concrete durability, the zoning indexes are determined as annual freeze-thaw cycles, annual precipitation, average temperature in July, and annual acid rainfall.

(3) Fuzzy clustering analysis and rough set theory were introduced into the environmental zonation for concrete durability, and the weights of durability zoning indexes were calculated. Based on this, Shaanxi Province was zoned into three first-level zones, namely the Freeze-Thaw Cycle

Zone, the Neutralization–Freeze-Thaw Interaction Zone, and the Neutralization Zone, and these three zones were then subdivided into nine second-level zones. The main reasons for durability deterioration of concrete structures were analyzed according to the environmental characteristics of each zone in the zoning map.

The method proposed in this paper can improve the quality and accuracy of environmental zonation. It can scientifically classify the environmental conditions of specific provinces, cities, or regions, and demonstrate the influence of durability factors in different regions. The zoning results can provide a scientific basis for durability design and maintenance of reinforced concrete structures.

Author Contributions: Conceptualization: D.L. and Y.W.; methodology: D.L. and Y.W.; software: D.L. and Y.W.; validation: D.L. and Y.W.; formal analysis: D.L. and S.Z.; investigation: S.Z.; resources: D.N.; data curation: D.L., Y.W., and S.Z.; writing—original draft preparation: D.L.; writing—review and editing: D.N.; visualization: D.L. and Y.W.; supervision: D.N.; project administration: D.L. and Y.W.; funding acquisition: D.L. and Y.W. All authors have read and agreed to the published version of the manuscript.

Funding: This work was financially supported by the National Natural Science Foundation of China (Grant No. 51808438 and No. 51878549) and by the Independent Research and Development project of the State Key Laboratory of Green Buildings in Western China (Grant No. LSZZ202016).

Acknowledgments: This work presented herein was conducted in the State Key Laboratory of Green Buildings in Western China at Xi'an University of Architecture & Technology. The authors gratefully acknowledge the support that has made this laboratory and its operation possible.

Conflicts of Interest: The authors declare that there is no conflict of interest.

References

1. Andersson, D.E.; Andersson, A.E. Sustainability and the Built Environment: The Role of Durability. *Sustainability-Basel* **2019**, *11*, 4926. [CrossRef]
2. Ashraf, W. Carbonation of cement-based materials: Challenges and opportunities. *Constr. Build. Mater.* **2016**, *120*, 558–570. [CrossRef]
3. Liu, H.; Luo, G.; Wang, L.; Gong, Y. Strength Time-Varying and Freeze-Thaw Durability of Sustainable Pervious Concrete Pavement Material Containing Waste Fly Ash. *Sustainability-Basel* **2019**, *11*, 176. [CrossRef]
4. Kessler, S.; Thiel, C.; Grosse, C.U.; Gehlen, C. Effect of freeze-thaw damage on chloride ingress into concrete. *Mater. Struct.* **2017**, *50*, 121. [CrossRef]
5. Gong, F.; Maekawa, K. Multi-scale simulation of freeze-thaw damage to RC column and its restoring force characteristics. *Eng. Struct.* **2018**, *156*, 522–536. [CrossRef]
6. Chen, M.; Wang, K.; Xie, L. Deterioration mechanism of cementitious materials under acid rain attack. *Eng Fail Anal* **2013**, *27*, 272–285. [CrossRef]
7. Kim, T.H.; Chae, C.U. Environmental impact analysis of acidification and eutrophication due to emissions from the production of concrete. *Sustainability-Basel* **2016**, *8*, 578. [CrossRef]
8. Zhang, P.; Wittmann, F.H.; Vogel, M.; Müller, H.S.; Zhao, T. Influence of freeze-thaw cycles on capillary absorption and chloride penetration into concrete. *Cem. Concr. Res.* **2017**, *100*, 60–67. [CrossRef]
9. Otieno, M.; Beushausen, H.; Alexander, M. Chloride-induced corrosion of steel in cracked concrete—Part I: Experimental studies under accelerated and natural marine environments. *Cem. Concr. Res.* **2016**, *79*, 373–385. [CrossRef]
10. Cheng, Y.; Zhang, Y.; Tan, G.; Jiao, Y. Effect of Crack on Durability of RC Material under the Chloride Aggressive Environment. *Sustainability-Basel* **2018**, *10*, 430. [CrossRef]
11. Müllauer, W.; Beddoe, R.E.; Heinz, D. Sulfate attack expansion mechanisms. *Cem. Concr. Res.* **2013**, *52*, 208–215. [CrossRef]
12. Ortega, J.M.; Esteban, M.D.; Williams, M.; Sánchez, I.; Climent, M.Á. Short-term performance of sustainable silica fume mortars exposed to sulfate attack. *Sustainability-Basel* **2018**, *10*, 2517. [CrossRef]
13. Ministry of Construction of the PRC. *GB 50010-2010, Code for Design of Concrete Structures*; China Architecture & Building Press: Beijing, China, 2015.
14. Ministry of Construction of the PRC. *GB/T50476-2019. Standard for Design of Concrete Structure Durability*; China Architecture & Building Press: Beijing, China, 2019.

15. European Committee for Standardization (CEN). *EN 1992-1-1, Euro-code 2: Design for Concrete Structures-Part 1-1: General Rules and Rules for Buildings*; CEN: London, UK, 2004.
16. Jin, W.; Lv, Q. Durability zonation standard of concrete structure design. *J. Southeast Univ. Engl. Ed.* **2007**, *23*, 98–104.
17. Wu, H.; Jin, W.; Yan, Y.; Xia, J. Environmental zonation and life prediction of concrete in frost environments. *J. Zhejiang Univ. Eng. Sci.* **2012**, *46*, 650–657.
18. Seo, S.; Lee, B.; Won, J. Comparative Analysis of Economic Impacts of Sustainable Vertical Extension Methods for Existing Underground Spaces. *Sustainability-Basel* **2020**, *12*, 975. [CrossRef]
19. Li, F.; Zhou, M.; Hu, M. Climate change in different geographical units and its impact on land production potential: A case study of Shaanxi Province, China. *Env. Sci. Pollut. Res.* **2019**, *26*, 22273–22283.
20. National Meteorological Information Center. China Meteorological Data Sharing Service System. Available online: http://data.cma.cn/ (accessed on 25 February 2020).
21. Compilation Committee of Local Chronicles of Shaanxi Province. Shaanxi Statistical Yearbook. Available online: http://www.sxsdq.cn/sqzlk/sxnj/ (accessed on 25 February 2020).
22. Papadakis, V.G.; Vayenas, C.G.; Fardis, M.N. Experimental investigation and mathematical modeling of the concrete carbonation problem. *Chem. Eng. Sci.* **1991**, *46*, 1333–1338. [CrossRef]
23. Chen, T.; Kuschner, W.G.; Gokhale, J.; Shofer, S. Outdoor air pollution: Nitrogen dioxide, sulfur dioxide, and carbon monoxide health effects. *Am. J. Med Sci.* **2007**, *333*, 249–256. [CrossRef]
24. Khan, R.R.; Siddiqui, M. Review on effects of particulates; sulfur dioxide and nitrogen dioxide on human health. *Int. Res. J. Env. Sci.* **2014**, *3*, 70–73.
25. Burns, D.A.; Aherne, J.; Gay, D.A.; Lehmann, C.M. Acid rain and its environmental effects: Recent scientific advances. *Atmos. Environ.* **2016**, *146*, 1–4. [CrossRef]
26. Müller, H.S.; Haist, M.; Vogel, M. Assessment of the sustainability potential of concrete and concrete structures considering their environmental impact, performance and lifetime. *Constr. Build. Mater.* **2014**, *67*, 321–337. [CrossRef]
27. Hooton, R.D.; Bickley, J.A. Design for durability: The key to improving concrete sustainability. *Constr. Build. Mater.* **2014**, *67*, 422–430. [CrossRef]
28. Alexander, M.; Bentur, A.; Mindess, S. Durability of Concrete: Design and Construction. CRC Press Taylor& Francis Group: Boca Raton, FA, USA, 2017; Volume 20, pp. 1–324.
29. Menz, F.C.; Seip, H.M. Acid rain in Europe and the United States: An update. *Env. Sci. Policy* **2004**, *7*, 253–265. [CrossRef]
30. Timm, H.; Borgelt, C.; Döring, C.; Kruse, R. An extension to possibilistic fuzzy cluster analysis. *Fuzzy Set Syst.* **2004**, *147*, 3–16. [CrossRef]
31. Xie, J.; Ping, C. *Fuzzy Mathematics Method and Applications*, 4th ed.; Huazhong University of Science & Technology Press: Wuhan, China, 2013; pp. 81–118.
32. Miyamoto, S.; Ichihashi, H.; Honda, K.; Ichihashi, H. *Algorithms for Fuzzy Clustering*; Springer, 2008; pp. 9–12.
33. Pawlak, Z. Rough sets and intelligent data analysis. *Inf. Sci.* **2002**, *147*, 1–12. [CrossRef]
34. Pawlak, Z. Rough set theory and its applications to data analysis. *Cybern. Syst.* **1998**, *29*, 661–688. [CrossRef]
35. Swiniarski, R.W.; Skowron, A. Rough set methods in feature selection and recognition. *Pattern Recogn. Lett.* **2003**, *24*, 833–849. [CrossRef]
36. Yao, Y.; Zhao, Y. Attribute reduction in decision-theoretic rough set models. *Inf. Sci.* **2008**, *178*, 3356–3373. [CrossRef]
37. Ministry of Construction of the PRC. *GB/T 51355-2019, Standard for Durability Assessment of Existing Concrete Structures*; China Building Industry Press: Beijing, China, 2019.
38. Sjors, H. The background: Geology, climate and zonation. *Acta Phytogeogr. Suec.* **1999**, *84*, 5–14.
39. Alexander, M.G.; Ballim, Y.; Stanish, K. A framework for use of durability indexes in performance-based design and specifications for reinforced concrete structures. *Mater. Struct.* **2008**, *41*, 921–936. [CrossRef]

sustainability

MDPI

Article

A Novel Construction Technology for Self-Anchored Suspension Bridge Considering Safety and Sustainability Performance

Xiaoming Wang [1], Xudong Wang [2], You Dong [3,*] and Chengshu Wang [4]

1 Key Laboratory for Bridge and Tunnel of Shannxi Province, Chang'an University, Xi'an 710064, China
2 Key Laboratory of Concrete and Prestressed Concrete Structures of Ministry of Education, Southeast University, Nanjing 210000, China
3 Department of Civil and Environmental Engineering, The Hong Kong Polytechnic University, Hong Kong 999077, China
4 Zhejiang Provincial Transport Planning, Design and Research Institute, Hangzhou 310000, China
* Correspondence: you.dong@polyu.edu.hk

Received: 20 March 2020; Accepted: 6 April 2020; Published: 8 April 2020

Abstract: To promote sustainable development of civil infrastructures, minimizing environmental impact and mobility disruptions have been elevated to a higher priority during decision-making for bridge construction scheme. This study presents a novel temporary pylon-anchor (TPA) technology for construction of self-anchored suspension bridges by considering not only safety performance, but also environmental impacts. A practical assessment method and index of sustainability associated with bridge construction technology are established to facilitate the selection of construction schemes. The sustainability index takes the environmental impact, traffic disruption, onsite construction materials and equipment, onsite construction cost, and onsite construction risk into consideration. The sustainability index associated with both conventional and novel construction methods is assessed and compared in this paper. Specifically, a novel girder-pylon antithrust system (GPAS) is proposed, which is the crucial component of the TPA technology in engineering application. In addition, an analytical approach is developed, considering both global load-carrying capacity and local stress distribution within the design and construction of the GPAS. The applicability and rationality of the proposed construction technology are illustrated by the successful application in real-world engineering. The field tests and sustainability assessment during the construction stage reveal that the proposed sustainability assessment method and analytical approach can facilitate the implementation of sustainable construction for self-anchored suspension bridges by considering both construction safety and sustainability.

Keywords: Sustainable construction; environmental impact and traffic disruptions; self-anchored suspension bridge; design-oriented analytical approach

1. Introduction

The onsite bridge construction activities can have significant impacts on environment, mobility, and safety [1,2]. The direct and indirect loss of environment and traffic disruptions resulting from the bridge construction can exceed the actual cost of the structure itself [3]. For instance, full-lane closures in large urban centers, or on highways or waterways with heavy traffic volumes, can have a significant economic impact on commercial and industrial activities in the region [4,5]; partial lane closures that occur alongside adjacent traffic can also lead to safety and environmental issues (e.g., extra CO_2 emissions due to traffic detour) [6]. Sustainable construction emphasizes an efficient use of natural resources to minimize the impacts of the built environment on the Earth and enhancing the quality of surrounding environment [7,8]. Because of the potential economic and safety impacts, minimizing environment and traffic disruptions is a goal that has been elevated to a higher priority when determining bridge sustainable construction scheme [9,10].

Self-anchored suspension bridges have found increasingly wide applications in urban bridge engineering as a landmark, due to its attractive architectural appearance and lightweight configuration [11–13]. Due to being anchored to the girder, the main cables could not be erected until the girder has been built on lots of temporary supports, which inevitably causes serious impacts on environment and traffic. This issue has emerged during the erection of several typical self-anchored suspension bridges, such as the Yongjong Bridge in Korea [13], San Francisco–Oakland Bay Bridge in the States [14], and Qingdao Bay Bridge in China [15]. Several new technologies were developed on the basis of the concept of sustainable construction to solve these issues [16,17], such as temporary earth-anchor method [18], temporary stayed cables method [19,20], and temporary compressive strut method [21]. Although these technologies could reduce mobility impacts, the use of expensive temporary structures (e.g., stayed cables, earth anchor, and compressive strut) increases environmental impact and onsite construction cost. Hence, to promote the construction sustainability for self-anchored suspension bridge, further research needs to focus on the development of the sustainable construction through new technology and new design-oriented analytical approaches.

A practice-oriented assessment method and index of sustainability on bridge construction technology are essential for selection of sustainable bridge construction schemes [22]. Penadés-Plà et al. [23] analyzed the life-cycle environmental impact of a prestressed concrete precast bridge from the economic point of view, and proposed an optimization-life-cycle assessment method. Chang et al. [24] established an index system and evaluated the sustainability of high-speed railway construction projects. Seo et al. [25] analyzed the economic impacts of three sustainable vertical extension methods for existing underground spaces. Although these methods could assess the construction sustainability, the real-world application of decision making for bridge construction schemes by considering sustainability is relatively scarce, and more studies are needed on this aspect. Therefore, it is urgent to establish a practical assessment method and index of sustainability on construction technology for urban bridge construction, which is addressed in this paper.

On the basis of the aforementioned studies, this study presents a novel temporary pylon-anchor (TPA) technology to promote sustainable construction, which is suited for self-anchored suspension bridges with the mid-span less than 300 m, three spans and two pylons. A practical assessment method and index of sustainability on construction technologies are established to facilitate making decisions for sustainable construction scheme. A novel girder-pylon antithrust system (GPAS) is proposed, which is the crucial component of the TPA technology in engineering application. For the reliable and cost-effective design of GPAS, a design-oriented two-phase framework is developed considering global load-carrying capacity and local stress distribution. In phase I, the global shear capacity and stiffness are designed through a set of specially derived practical formulas, which capture the main characteristics of the slip and uplift behavior at steel-concrete joint surface within an antithrust system. In phase II, the local stress distribution is improved based on the effects of different parameters induced by grouped parametric analyses using 3D elaborate finite element analysis. The applicability and rationality of the sustainable construction new technology are illustrated by the first successful application in real-world engineering. The field tests and construction sustainability assessment show that the proposed sustainability assessment method and design-oriented analytical approach facilitate the implementation of sustainable construction for self-anchored suspension bridge.

2. Sustainability Assessment on Construction Technologies for Self-Anchored Suspension Bridge

2.1. Environmental Impact and Traffic Disruption Caused by Traditional Construction Technologies

As shown in Figure 1a, the traditional construction technology needs lots of temporary supports to build girders, which inevitably causes serious environmental impact and traffic disruption.

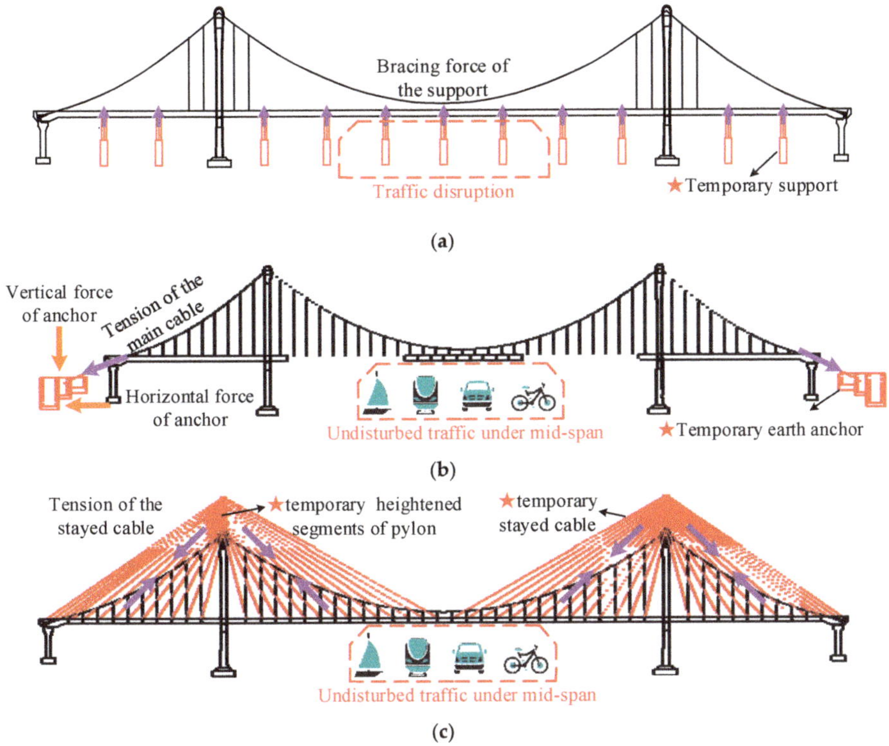

Figure 1. Traditional construction technologies for self-anchored suspension bridge: (**a**) Temporary supports technology [13,26]; (**b**) temporary earth-anchor technology [18]; and (**c**) temporary stayed-cables technology [19,20].

As shown in Figure 1b, for reducing traffic disruptions, the temporary earth-anchor method builds temporary earth anchor blocks to enable a construction sequence that is similar to a conventional suspension bridge. The technology has been applied in the Zhuyuan Bridge (33 m + 90 m + 33 m) in China [18]. However, the building of earth-anchor also damages the surrounding environment and produces pollution.

As shown in Figure 1c, to avoid traffic disruptions, the temporary stayed cables technology uses temporary stayed-cables to erect girder segments, just as a cable-stayed bridge, and these stayed-cables aren't removed until the main cables and hangers are erected. The technology has been applied in the E'gongyan Rail Bridge (210 m + 600 m + 210 m) in China [19], and the Duisburg Bridge [20]. Although the technology reduces both environmental impact and traffic disruption, the use of expensive temporary structures (lots of stayed cables and heightened segments of the pylon) increases onsite construction cost and time.

2.2. Novel TPA Technology to Improve Construction Sustainability

In this study, a novel temporary pylon-anchor (TPA) technology is proposed to solve the issue in a safe and cost-effective manner for minimizing environment and traffic disruptions. As shown in Figure 2, the horizontal cable force is transferred to the pylon through the side span girder and is resisted by means of the bending bearing capacity of the pylon. Thus, the mid-span girder is lifted in sections and connected to the hangers. After the entire girder is erected, the horizontal cable force is transferred from the pylon to the girder, and the structure is transformed into a permanent self-balanced system. In this way, the permanent environmental impact and traffic disruption are eliminated during construction, and the onsite construction cost, time, and risk are minimized.

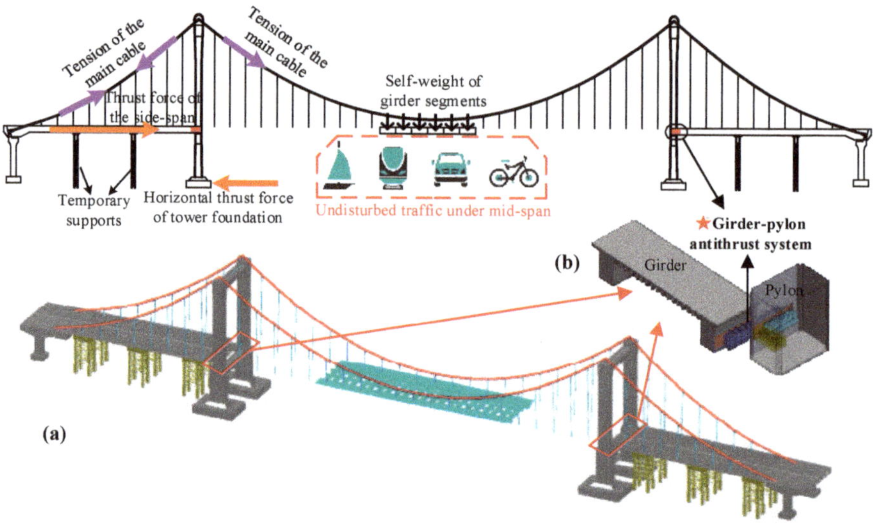

Figure 2. Temporary pylon-anchor technology: (**a**) Construction process; (**b**) Gider-pylon antithrust system.

2.3. Primary Sustainability Performance of Construction Technologies

As shown in Table 1, the traditional temporary supports technology results in not only serious environmental impact and traffic disruption, but also high onsite construction cost, time, and risk. Even if the temporary earth-anchor technology and stayed-cables technology reduce traffic disruption, the other important sustainable performance still needs further improvement.

Table 1. Construction sustainability evaluation of construction technologies for self-anchored suspension bridge.

Performance \ Technology	**Temporary Supports** [13,26]	**Temporary Earth-Anchor** [18]	**Temporary Stayed-Cables** [19,20]	**The Proposed Temporary Pylon-Anchor**
Environmental impact	**Serious:** environmental damage caused by temporary supports' foundation	**Serious:** environmental damage caused by temporary earth-anchorage	**Almost none**	**Almost none**
Traffic disruption	**Serious:** traffic under the bridge is blocked by temporary supports	**Almost none**	**Almost none**	**Almost none**
Onsite construction materials and equipment	**High:** lots of temporary supports	**High:** expensive temporary earth-anchorage	**High:** lots of temporary stayed-cables, heightened segments of pylon	**Low:** cheap girder–pylon antithrust system

Table 1. *Cont.*

Technology / Performance	Temporary Supports [13,26]	Temporary Earth-Anchor [18]	Temporary Stayed-Cables [19,20]	The Proposed Temporary Pylon-Anchor
Onsite construction time	**Long:** lots of time-consuming for erection and removal of supports	**Long:** lots of time-consuming for erection and removal of earth-anchorage	**Long:** lots of time-consuming for erection and removal of stay-cables and heightened segments of pylon	**Short:** Fast erection and removal of girder–pylon antithrust system
Onsite construction risk	**High:** temporary supports subjected to flood and collision of vehicles	Low	Low	Low

2.4. Assessment Method and Index Considering Construction Sustainability

In order to facilitate decision-making for sustainable bridge construction scheme, a practical assessment procedure is proposed as shown in Figure 3. The detailed procedure is explained as follows:

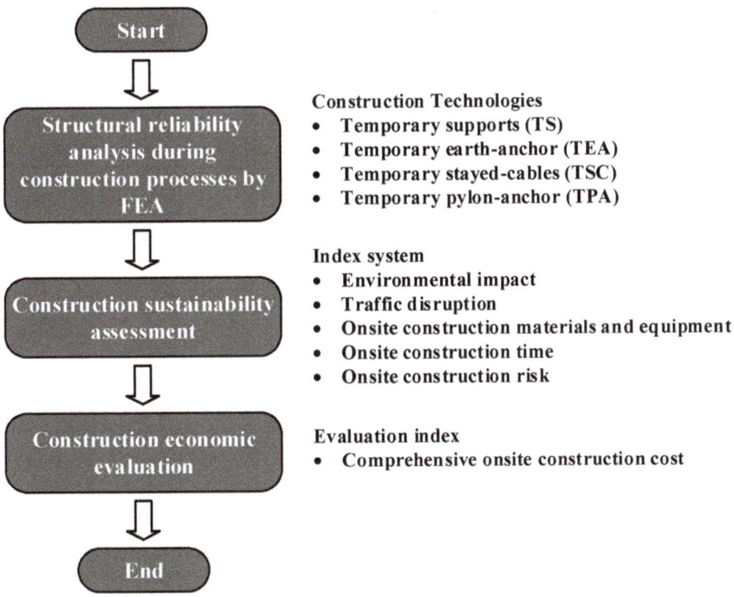

Figure 3. Practice-oriented assessment procedure of sustainability assessment of bridge construction technology.

(1) Structural safety analysis

To ensure the structural safety during the entire construction processes, the global step-by-step forward model and local 3D solid elaborate model should be analyzed for each construction technology, respectively. The analytical approach proposed in Section 3 presents the detailed analysis procedure and mechanical performance. The structural analysis performance includes the strength, rigidity, stability, etc.

Through this step, those construction technologies that cannot meet the safety requirement are excluded.

(2) Construction sustainability assessment

Considering the various influencing factors on construction sustainability, this paper uses an index system by considering the following aspects: Environmental impact, traffic disruption, onsite construction materials and equipment, onsite construction cost, and onsite construction risk. Then, the hierarchical evaluation method based on a rose chart is employed to compare sustainability associated with different construction schemes. As shown in Figure 4, the area S_i enclosed by connecting lines is taken as analysis index of the sustainability associated with the construction technology. As indicated, the indices of four technologies are indicated in Figure 4 for illustrative purpose, and the TPA technology has comprehensive performance on construction sustainability.

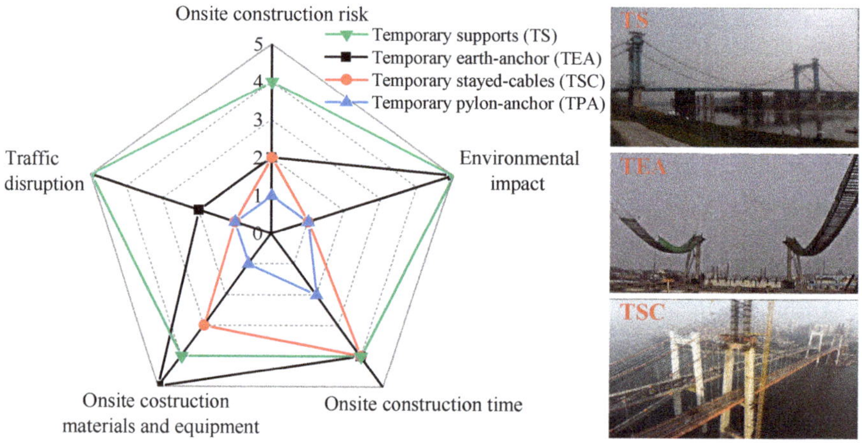

Figure 4. Construction sustainability assessment.

Through step 2, the construction technology with the best sustainability is efficiently screened out.

(3) Construction economic evaluation

In this step, the direct and indirect construction cost of the most sustainable construction technology is assessed. The comprehensive onsite construction cost is the sum of C_1, C_2, C_3, C_4, and C_5. C_1 is direct and indirect losses due to environmental impact; C_2 is the direct and indirect losses caused by traffic disruption; C_3 is the cost of materials and equipment in onsite construction; C_4 is the cost of onsite construction time, which is obtained by multiplying labor cost (per day) and days; and C_5 represents the cost of onsite construction risks, which is quantified by multiplying insurance amount and failure probability.

3. Design-Oriented Analytical Approach for Novel Antithrust System

As shown in Figure 2, the GPAS is the crucial component of the proposed TPA technology. In the GPAS, the horizontal cable force is smoothly transferred from the side span girder to the pylon through the thrust shoulder. The shear capacity and stiffness of the thrust shoulder need to be designed to meet the construction performance requirements, and the girder and pylon should be nondestructive.

3.1. *Novel Girder–Pylon Antithrust System*

The GPAS can be classified on the basis of the gap width between the side span girder and the pylon as follows: Type I (Figure 5a) for the wide gap, and Type II (Figure 5b) for the narrow gap. A concrete side span girder is usually installed in a self-anchored suspension bridge to balance the uplift component of cable force. Hence, grouped perfobond rib connectors [Perfobond Leiste in

German (PBL)] and U-shaped anchor rebars are used to fasten the antithrust system to the lateral side of the side span girder.

(1) Type I

Figure 5a shows the Type I antithrust system, consisting of a girder-side thrust shoulder, a pylon-side thrust shoulder, and a force transmission brace located between the set of thrust shoulders. Each thrust shoulder is composed of a grille frame, grouped PBLs, and U-shaped anchor rebars that are welded to the two sides of the wallboard. The grille frame comprises longitudinal, vertical, and transverse plates. The thrust shoulder is fastened to the lateral side of the side span girder and pylon with the aid of grouped PBLs and U-shaped anchor rebars that prevent the thrust shoulders from slip and uplift.

(2) Type II

The only difference between Types II (Figure 5b) and I is that the pylon-side thrust shoulder is replaced by the foot brace to effectively transfer the thrust force and disperse the narrow gap stress.

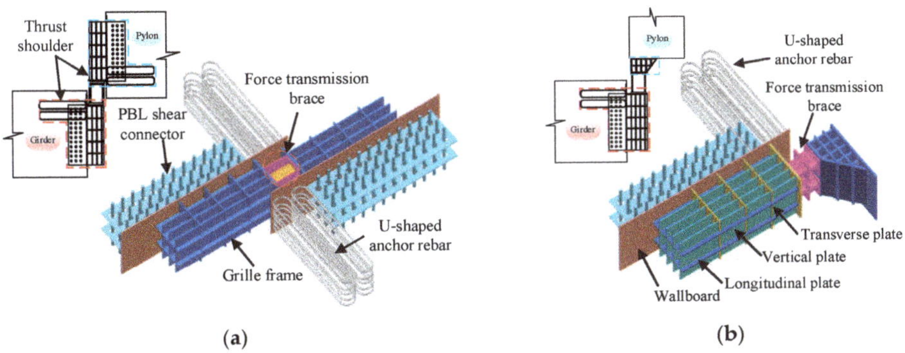

Figure 5. Schematic diagram of GPAS (Girder-Pylon Antithrust System): (**a**) Type I antithrust system and (**b**) Type II antithrust system.

3.2. Failure Modes and Mechanism of Force Transmission

The three failure modes of the thrust are considered to meet the construction performance requirements. (I) Shear slip of PBL shear connector [27–29]: The shear slip deformation of PBL connector is limited to ensure that the girder and pylon are nondestructive. The increase in shear capacity after sliding is excluded in the design; (II) uplift of wallboard: Under the strong thrust force, the out-of-plane deformation of the wallboard at the head position must be avoided; and (III) buckling failure of grille frame: The grille frame must have sufficient rigidity in the 3D space. The overall or local buckling of the grille frame must be avoided. No cyclic loading occurs although the structural stress varies seriously during construction. Hence, the fatigue failure of the thrust shoulder can be ignored.

As shown in Figure 6, the horizontal cable force is transmitted to the wallboard, U-shaped anchor rebars, and PBL shear connectors through the grille frame to achieve the force transmission between the girder and pylon. The shear capacity of PBL shear connectors is the chief resistance for the thrust force. The eccentric effect of the grille frame would cause the wallboard uplift at the head position, enabling the U-shaped anchor rebars and PBL shear connectors to withstand the transverse pull-out force for resisting the out-of-plane deformation.

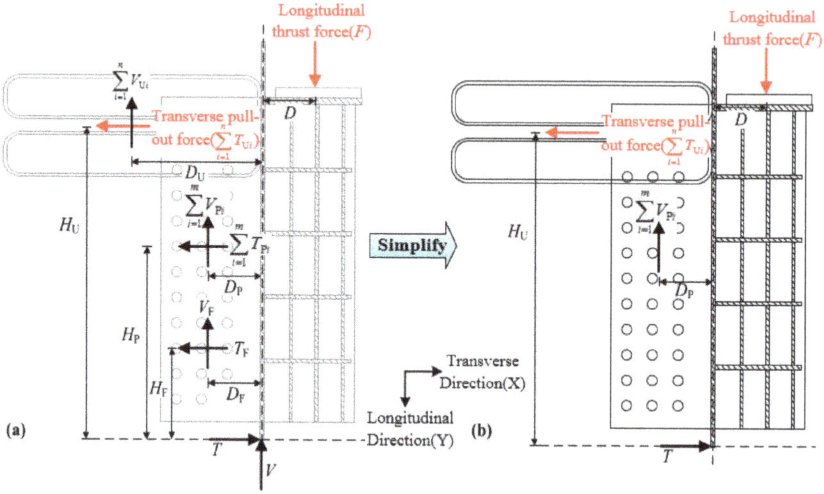

Figure 6. Mechanism diagram of force transmission of thrust shoulder: (**a**) Calculation scheme; (**b**) Simplified calculation scheme.

The equilibrium formula is established on the basis of the calculation scheme (Figure 6a) as follows:

$$\left.\begin{array}{l} F=\sum\limits_{i=1}^{n} V_{\mathrm{U}i}+\sum\limits_{i=1}^{m} V_{\mathrm{P}i}+V_{\mathrm{F}}+V \\ T=\sum\limits_{i=1}^{n} T_{\mathrm{U}i}+\sum\limits_{i=1}^{m} T_{\mathrm{P}i}+T_{\mathrm{F}} \\ F\cdot D=H_{\mathrm{U}}\cdot\sum\limits_{i=1}^{n} T_{\mathrm{U}i}-D_{\mathrm{U}}\cdot\sum\limits_{i=1}^{n} V_{\mathrm{U}i}+H_{\mathrm{P}}\cdot\sum\limits_{i=1}^{m} T_{\mathrm{P}i}-D_{\mathrm{P}}\cdot\sum\limits_{i=1}^{m} V_{\mathrm{P}i}+H_{\mathrm{F}}\cdot T_{\mathrm{F}}-D_{\mathrm{F}}\cdot V_{\mathrm{F}} \end{array}\right\} \quad (1)$$

where F is the longitudinal thrust force, D is the distance of the longitudinal thrust force from the center of rotation in the X direction, $\sum\limits_{i=1}^{n} V_{Ui}$ is the total force of U-shaped rebars in the Y direction, $\sum\limits_{i=1}^{n} T_{Ui}$ is the transverse (X-direction) pull-out force of U-shaped rebars, n is the number of U-shaped rebars, $\sum\limits_{i=1}^{m} V_{Pi}$ and $\sum\limits_{i=1}^{m} T_{Pi}$ are the total forces of penetrating rebar in the Y and X directions, m is the number of penetrating rebars, V_F and T_F are the friction between the steel and concrete in the Y and X directions, V and T are the reactions at the end of the wallboard in the Y and X directions, H_U and D_U are the distances of the total force of U-shaped rebars from the center of rotation in the Y and X directions, H_P and D_P are the distances of the total force of penetrating rebar from the center of rotation in the Y and X directions, and H_F and D_F are the distances of friction from the center of rotation in the Y and X directions, respectively.

The following simplifications are adopted to ensure design reliability and improve practicability: The thrust force is only borne by PBL shear connectors. The internal force redistribution caused by slip deformation is ignored, indicating that the relative increase in shear capacity after sliding can be ignored [30]; the friction and embedment effects of the wallboard on shear resistance are ignored; and the out-of-plane uplift of the wallboard is significant at the head zone, making it reasonable to assume that the transverse pull-out force is only borne by the U-shaped anchor rebars. Based on the above simplifications (Figure 6b), Formula (1) can be simplified to Formula (2).

$$\left.\begin{array}{l} F=\sum\limits_{i=1}^{m} V_{\mathrm{P}i} \\ T=\sum\limits_{i=1}^{n} V_{\mathrm{U}i} \\ F\cdot D=H_{\mathrm{U}}\cdot\sum\limits_{i=1}^{n} T_{\mathrm{U}i}-D_{\mathrm{P}}\cdot\sum\limits_{i=1}^{m} V_{\mathrm{P}i} \end{array}\right\} \tag{2}$$

3.3. Global Design Formulas

The calculation formula of PBL bearing capacity (V_{pud}) provided by the design code [31] is adopted to improve the applicability of design formula. The shear heterogeneity coefficient of PBL shear connector $\alpha = 0.4$ is considered for security redundancy [32]. The formula to estimate the number of PBL shear connectors is derived as follows:

$$N_{\mathrm{P}} \geq \frac{F}{\alpha\cdot V_{\mathrm{pud}}}=\frac{F}{0.4\cdot(1.4\cdot(d^2-d_{\mathrm{s}}^2)f_{\mathrm{cd}}+1.2d_{\mathrm{s}}^2 f_{\mathrm{sd}})} \tag{3}$$

where N_{P} is the number of PBL shear connectors, F is the longitudinal thrust force in Formula (1), α is the shear heterogeneity coefficient, V_{pud} is the shear capacity of PBL shear connector under the ultimate limit state, d is the diameter of circular hole in the steel plate, d_s is the diameter of penetrating rebar, f_{cd} is the design value of concrete axial compressive strength, and f_{sd} is the design value of penetrating rebar tensile strength.

The anchorage length of U-shaped anchor rebar must reach the yield strength before anchorage failure to meet the construction performance requirements. The design code [33] indicates that the minimum anchorage length of HRB400 rebar adopted in this paper should be greater than 25 times the rebar diameter. The formula to estimate the number of U-shaped anchor rebars is derived as follows:

$$N_{\mathrm{U}} \geq \frac{F\cdot D+\sum\limits_{i=1}^{m}(D_{\mathrm{P}i}\cdot V_{\mathrm{pud}})}{2\cdot H_{\mathrm{U}}\cdot A_{\mathrm{s}}\cdot f_{\mathrm{sd}}} \tag{4}$$

where N_{U} is the number of U-shaped anchor rebars, $D_{\mathrm{P}i}$ is the distance of the ith penetrating rebar from the center of rotation in the X direction, and A_s is the cross-sectional area of U-shaped anchor rebar.

The longitudinal plate of the grille frame is designed considering the rigidity criterion. The thickness and number of longitudinal plates must ensure that the structural deformation meets the construction requirements. The transverse and vertical plates of the grille frame are designed on the basis of stability control criterion, and their thickness and spacing are set in accordance with the code [34] to avoid buckling.

3.4. Design-Oriented Two-Phase Analytical Framework

As indicated in Figure 7, the design-oriented analytical framework of the girder-pylon thrust shoulder is constructed. The detailed steps are provided as follows:

1. Lectotype

The type of antithrust system is selected on the basis of specific structural parameters, such as the configuration and material of the side span girder and bridge pylon and the gap width between the girder and pylon.

2. Determination of unfavorable construction state

Global geometric nonlinear analysis of the entire construction process is conducted through the spatial cable and beam element finite element (FE) simulation, and the maximum design load of the GPAS is determined.

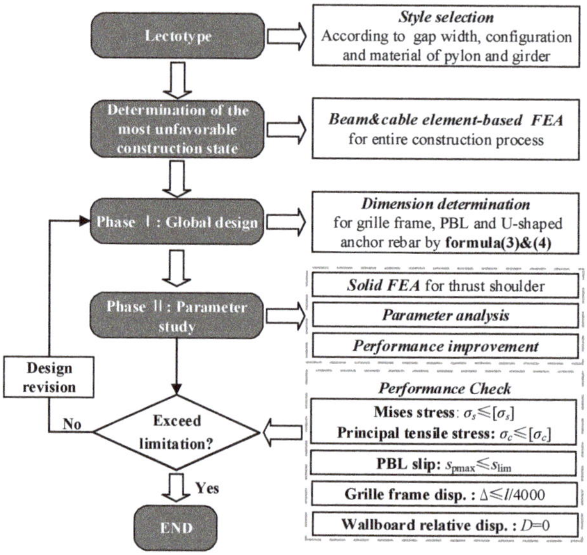

Figure 7. Design-oriented analytical framework for girder-pylon thrust shoulder.

3. Phase I: Global design

The numbers of PBL shear connectors and U-shaped anchor rebars are determined using Formulas (3) and (4) on the basis of the design criterion in Section 3.3. The grille frame is designed considering rigidity and stability control criteria.

4. Phase II: Parametric study

Grouped parametric analyses are performed to reveal the effects of parameter variation on structural performance and to improve the local performance. A high-fidelity 3D solid FE model is established to analyze the local effect and avoid stress distortion and size disturbance effect. The nonlinear surface contact behavior is simulated to consider the friction effect between the steel and concrete.

5. Performance check

The performance of the parametric study in step 4 is verified using the performance indicators. The design procedure is completed when the performance is verified, otherwise, the design is updated using steps 3-5. The performance indicators and relative thresholds are provided as follows:

(1) The principal tensile stress of concrete σ_c and Mises stress of steel σ_s must not exceed the limit value.

$$\begin{cases} \sigma_c \leq [\sigma_c] \\ \sigma_s \leq [\sigma_s] \end{cases} \tag{5}$$

where $[\sigma_c]$ is the allowable tensile stress of concrete, and $[\sigma_s]$ is the allowable yield stress of steel (Detailed reference values are shown in Table 2).

(2) The slip of PBL shear connectors must not exceed the limit value [31].

$$S_{\rm pmax} = \frac{0.6V_{\rm pud}}{29\sqrt{(d-d_{\rm s})d_{\rm s}f_{\rm ck}E_{\rm c}}} \leq s_{\rm lim} = 0.2 \text{ mm} \tag{6}$$

where $S_{\rm pmax}$ is the maximum slip of the joint surface, $s_{\rm lim}$ is the slip limit, $f_{\rm ck}$ is the standard value of concrete axial compressive strength, and $E_{\rm c}$ is the Young's modulus of concrete.

(3) The deformation of grille frame must not exceed the limit value [34].

$$\Delta \leq l/4000 \text{ mm} \tag{7}$$

where Δ and l are the longitudinal displacement and design length of the grille frame, respectively.

(4) No relative deformation is found between the wallboard and the girder, that is, relative deformation D= 0.

4. Design and Engineering Implementation

The Dongtiao River Bridge in Huzhou, China, which is a self-anchored cable-stayed-suspension bridge with a span of 75 m + 228 m + 75 m and a semi-floating structural system (Figure 8), is used as the illustrative example. The bridge tower is a steel structure. The steel-concrete composite girder is used in the mid-span to reduce the self-weight, and the stay cable and prestressed concrete girder are used in the side span to balance the thrust force transmitted from the main cable to the bridge tower. Four sets of GPAS are used in the bridge to balance the thrust force of the side span generated during the construction process using the proposed TPA technology. The type I GPAS is selected because the gap between the girder and the pylon is approximately 1 m.

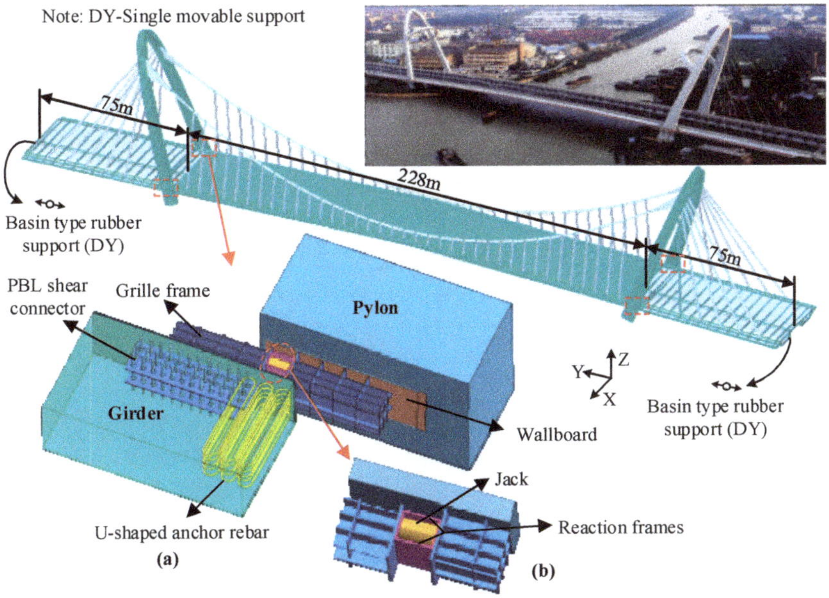

Figure 8. The global FE model of Dongtiao river bridge and the layout of GPAS: (**a**) Girder-pylon antithrust system and (**b**) force transmission brace.

4.1. Determination of Unfavorable Construction State

A temporary GPAS for the Dongtiao River Bridge is designed on the basis of the design procedure in Section 3.4. A 3D beam-link-cable FE model (Figure 8) is presented for the entire construction analysis of the structural system. Then, the construction stage analysis is conducted to obtain the maximum thrust force of the side span girder during construction, as shown in Figure 9. With the progress of construction, the maximum thrust force on the GPAS is 17,400 kN. The antithrust system is installed at the two sides of pylons, and a load safety factor of 1.4 [35] is considered on the basis of construction experience. The maximum thrust force on the unilateral antithrust system is 17,400/2 × 1.4 = 12,180 kN, which is defined as the most unfavorable thrust force.

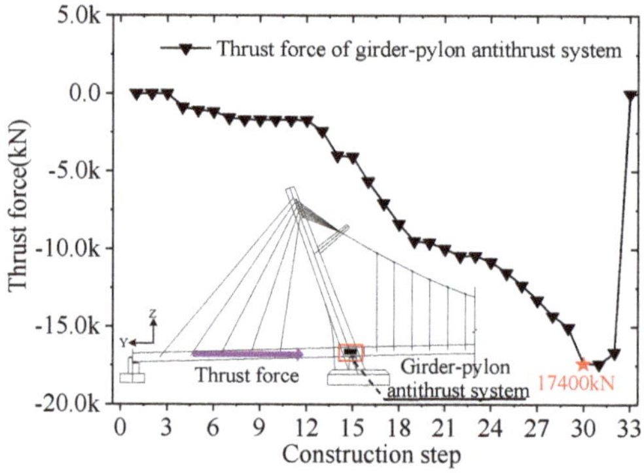

Figure 9. Thrust force variation of GPAS during the construction process.

4.2. Phase I: Global Design

The test results [36] indicated that the rebar diameter in PBL shear connectors ranges from 10 mm to 25 mm. The PBL shear connectors, with a 60 mm diameter hole and 25 mm diameter penetrating rebar in the hole, are arranged in 200 mm spacing along with longitudinal and transverse directions on the basis of the construction code requirements [31]. The number of PBL shear connectors is preliminarily calculated using Formula (8). The Twin-PBL shear connector [37] is adopted to increase the torsional resistance and overall stiffness of PBL shear connectors. Considering that the Twin-PBL shear connector is arranged in two-holes, 45-holes are arranged in 15-rows for each perforated rib, as shown in Figure 10a.

$$N_P \geq \frac{F}{\alpha \cdot V_{pud}} = \frac{F}{0.4 \cdot (1.4 \cdot (d^2 - d_s^2) f_{cd} + 1.2 d_s^2 f_{sd})} = \frac{12180 \text{ kN}}{136 \text{ kN}} \approx 90 \quad (8)$$

On the basis of the aforementioned design principle, the anchorage length of U-shaped anchor rebar should be longer than 25 d_u, where d_u is the U-shaped anchor rebar diameter, and the number of U-shaped anchor rebars should be calculated using Formula (9). As shown in Figure 10a, three groups of six U-shaped anchor rebars with 25 mm diameter and 1.8 m length are initially designed to meet safety redundancy. As shown in Figure 10b, the preliminary size of grille frame is proposed.

$$N_U \geq \frac{F \cdot D + \sum_{i=1}^{m} (D_{Pi} \cdot V_{pud})}{2 \cdot H_U \cdot A_s \cdot f_{sd}} = \frac{12180 \times 10^3 \times 0.615 + 6.885 \times 10^6}{2 \times 4.2 \times 4.9 \times 330 \times 10^2} \approx 11 \quad (9)$$

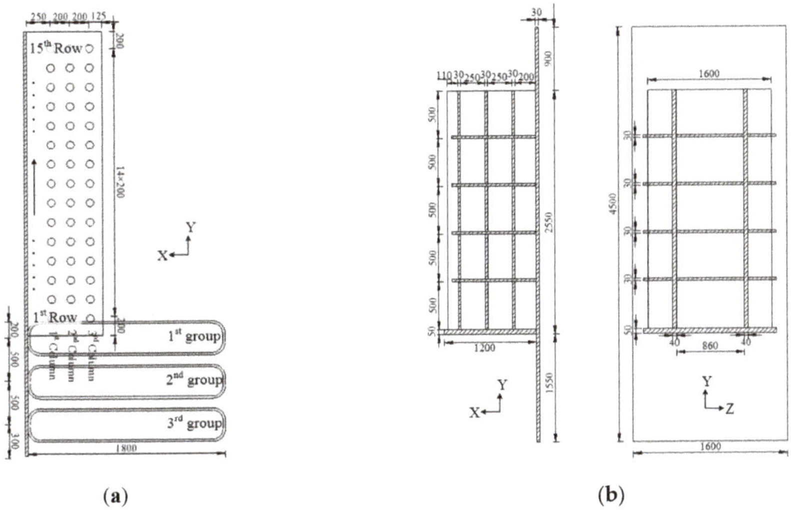

Figure 10. The layout of the thrust shoulder: (**a**) Grouped PBL shear connectors and U-shaped anchor rebars, and (**b**) grille frame.

4.3. Phase II: Parametric Study

4.3.1. FEA

The general software package ANSYS is used to simulate the behavior of the thrust shoulder (Figure 11). After considering the size effect and the stress distortion on the coupling nodes of different elements [38,39], the concrete in the anchorage zone is simulated by solid element SOLID65 [40]. The solid element SOLID92 is used to simulate the grille frame and PBL shear connectors, (including the perforated rib and penetrating rebar), and the beam element BEAM4 is used to simulate the U-shaped anchor rebars.

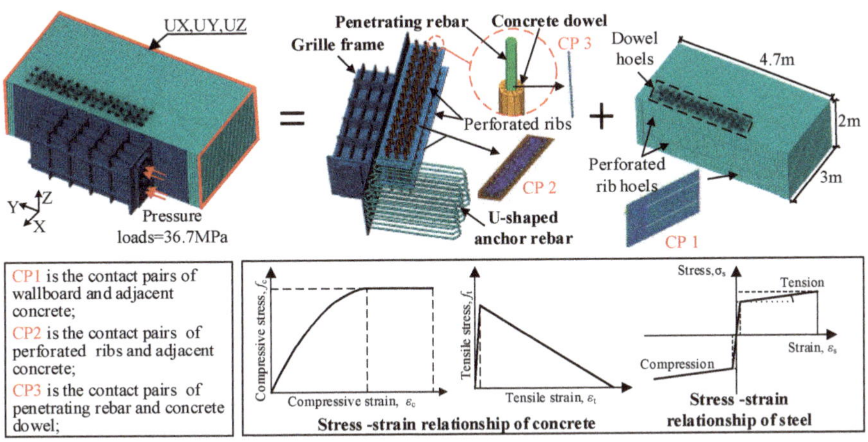

Figure 11. The high-fidelity 3D solid finite element model.

In order to simulate the friction between the steel and concrete (Figure 11), the nonlinear surface-to-surface contact elements CONTA174 and TARGE170 are used to simulate three contract pairs (CPs), namely, CP1 between the wallboard and concrete, CP2 between the perforated ribs and concrete, and CP3 between the penetrating rebar and concrete dowel. CONTA174 element is adopted to the concrete surface, whereas TARGE170 element is used to the steel plate surface. The tangential behavior is assumed via a penalty friction formulation with a friction coefficient of 0.35 [38].

The displacements in X, Y, and Z directions of the concrete anchorage zone are constrained in the model calculation. The unfavorable thrust force is 12,180 kN, and the direction is Y-axis. The thrust force is applied in the form of surface load to ensure the model load accuracy.

The material properties are shown in Table 2. The material of the grille frame, wallboard, and perforated rib is Q345C. The material of the penetrating and U-shaped anchor rebars is HRB400, and that of the concrete is C50. The nonlinear behavior of materials is considered to increase the accuracy of FE simulation. The uniaxial stress–strain relationship of concrete in compression is obtained using the Hongnestad model, and incorporated into the FE model with Multi-linear Isotropic Hardening (MISO) option. The linear model of concrete in tension is adopted for the relationship between stress and strain. After crack resistance strength, the tensile stress linearly reduces to zero toward the ultimate strain because of the softening of concrete in tension (Figure 11) [41]. The stress–strain relationship of steel components is regarded as elastic-perfectly-plastic, and incorporated into the FE model with Bilinear Kinematic Hardening (BKIN) option. The assumed stress–strain curve of steel is shown in Figure 11, which is assumed to be a bilinear curve, including the strain hardening effect on the tension and compression sides [41].

Table 2. Material properties.

Material	Grade	Compressive Strength/MPa	Tensile Strength/MPa	Yield Strength/MPa	Young's Modulus/GPa
Structural steel	Q345C	270	270	345	210
Rebar	HRB400	330	330	400	200
Concrete	C50	22.4	1.83	-	34.5

4.3.2. Parametric Analysis

Various FE models are used in the parametric analysis to examine the effects of penetrating rebar diameter (P_D), hole space on the perforated rib (P_S), diameter (U_D) and length (U_L) of the U-shaped anchor rebar, number of longitudinal plates (N_L), and number of transverse plates (N_T) of the grille frame. The test model T1 is established on the basis of the preliminary design, and all model parameters are listed in Table 3.

Table 3. Grouped models for parametric analyses.

Specimens	P_S (mm)	P_D (mm)	U_L (mm)	U_D (mm)	N_L	N_T
T1	200	25	1800	25	2	5
T1-P_D-20	200	**20**	1800	25	2	5
T1-P_D-22	200	**22**	1800	25	2	5
T1-P_D-28	200	**28**	1800	25	2	5
T1-P_D-32	200	**32**	1800	25	2	5
T1-P_S-100	**100**	25	1800	25	2	5
T1-P_S-150	**150**	25	1800	25	2	5
T1-P_S-250	**250**	25	1800	25	2	5
T1-P_S-300	**300**	25	1800	25	2	5
T1-U_D-20	200	25	1800	**20**	2	5
T1-U_D-22	200	25	1800	**22**	2	5
T1-U_D-28	200	25	1800	**28**	2	5
T1-U_D-32	200	25	1800	**32**	2	5
T1-U_L-1600	200	25	**1600**	25	2	5

Table 3. *Cont.*

Specimens	P_S (mm)	P_D (mm)	U_L (mm)	U_D (mm)	N_L	N_T
T1-U_L-2000	200	25	**2000**	25	2	5
T1-U_L-2200	200	25	**2200**	25	2	5
T1-U_L-2400	200	25	**2400**	25	2	5
T1-N_L(3)-N_T(5)	200	25	1800	25	**3**	5
T1-N_L(4)-N_T(5)	200	25	1800	25	**4**	5
T1-N_L(5)-N_T(5)	200	25	1800	25	**5**	5
T1-N_L(2)-N_T(4)	200	25	1800	25	2	**4**
T1-N_L(2)-N_T(6)	200	25	1800	25	2	**6**
T1-N_L(2)-N_T(7)	200	25	1800	25	2	**7**

Effects of Penetrating Rebar Diameter and Perforated Rib Hole Spacing

To investigate the effect of penetrating rebar diameter and perforated rib hole spacing on the yield stress of the penetrating rebar, five types of penetrating rebar diameter P_D and hole spacing P_S, as shown in Table 3, are considered, respectively.

Figure 12a,b illustrate the effects of diameter and hole spacing on Mises stress of the penetrating rebar at the first column, respectively. As shown in Figure 12a, the peak Mises stress appears in the 12th row position, thereby indicating that the middle PBL shear connectors bear a large thrust force. As shown in Figure 12b, the maximum stress value of the penetrating rebar is less than that of other spacing when the hole spacing is 100 mm, indicating that the spacing between holes is extremely short, and the PBL shear connectors do not exert the maximum effect. The peak stress appears and moves to the loading side when the hole spacing is greater than 200 mm, indicating that the PBL shear connectors exert the maximum effect under the structural size constraint. The results of FE analysis are consistent with the design formulas deduced in phase I, and the reliability of Formula (3) is verified.

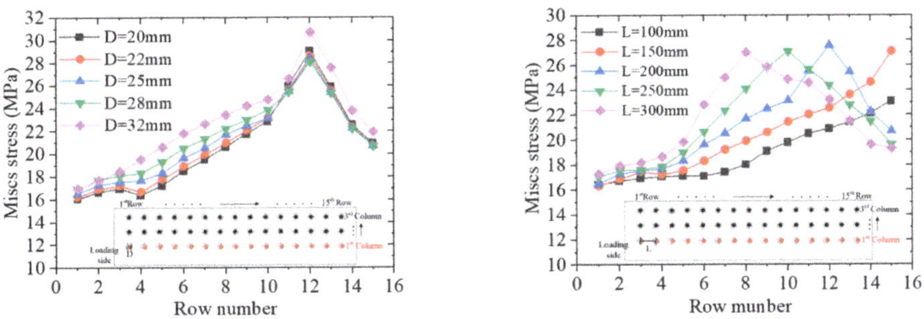

(**a**) (**b**)

Figure 12. Comparison of PBL performance with different parameters: (**a**) Effect of diameter on Mises stress of penetrating rebar at 1st column; (**b**) effect of hole spacing on Mises stress of penetrating rebar at 1st column.

Effects of the Diameter and Length of U-Shaped Anchor Rebar

Five types of diameter U_D and length U_L are considered to investigate the effects of the diameter and length of U-shaped anchor rebar on tensile stress, as shown in Table 3. Figure 13a shows the effect of diameter on the tensile stress of U-shaped anchor rebar. The maximum stress of the U-shaped anchor rebar gradually decreases with the increase in diameter. Figure 13b shows the effect of length on the tensile stress of U-shaped anchor rebar. The maximum tensile stress of the U-shaped anchor rebar slowly decreases with the increase in length. The maximum stress decreases 84 and 42 MPa when the diameter and length of U-shaped anchor rebar vary from 22 mm to 32 mm and from 1.6 m to 2.4 m, respectively, indicating that diameter plays a great role in the tension capacity of the U-shaped anchor rebar. This condition is mainly due to the fact that the maximum tensile stress appears in the weld between the U-shaped anchor rebar and the wallboard, and the increase in the diameter of U-shaped anchor rebar leads to the increase in the butt weld area.

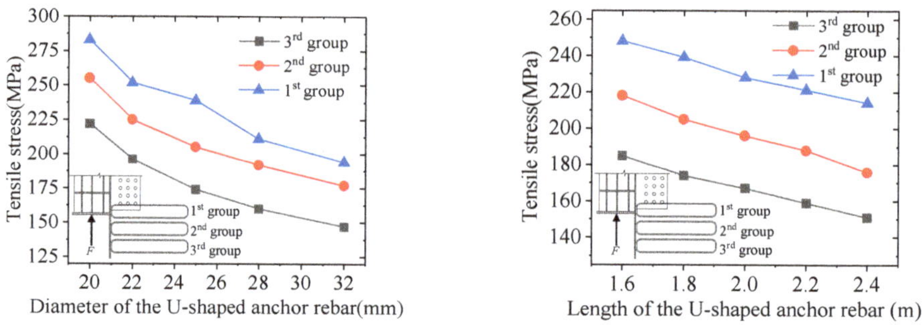

Figure 13. Comparison of tensile stress of U-shaped anchor bars in different (**a**) diameters and (**b**) lengths.

Effects of the Number of Grille Plates

The grille frame transmits the thrust force to the wallboard through several soldering seams connected between the longitudinal plates, transverse plates, and the wallboard. The effect of the number of different kinds of plates on the thrust-sharing proportion of the grille frame, as shown in Table 3, are considered to investigate the optimal number of each kind of plates.

The thrust-sharing proportion of the longitudinal plate exceeds 80%, and the transverse plates do not exceed 20%, as shown in Figure 14, indicating that the longitudinal plate is the important force transfer component. The thrust-sharing proportion transmitted by the two components does not significantly vary with the increase in the number of longitudinal and transverse plates.

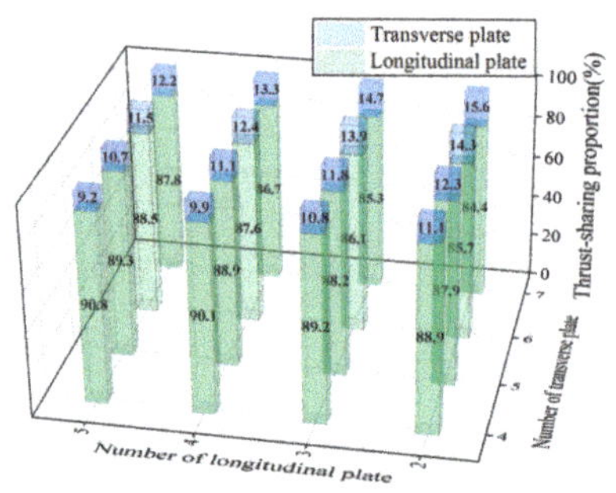

Figure 14. Effect of number of plates on thrust-sharing proportion.

4.3.3. Performance Improvement

In this section, the global design of phase I is improved on the basis of the aforementioned parametric analyses. The design of PBL and U-shaped anchor rebar is the key factor affecting the thrust shoulder performance. The stress is more sensitive to the change of hole spacing and diameter of U-shaped anchor rebar, so special attention should be paid to the selection of their sizes. However, the number of grille plates is not a key factor, and the change of the number has little effect on the thrust-sharing proportion of each component. Therefore, according to the results of parameter analysis and considering the safety redundancy of the structure, the improved parameters are as follows: (1) The penetrating rebar diameter is 25 mm, and the hole spacing is 200 mm; (2) the diameter of U-shaped anchor rebar is 28 mm, and the length is 2 m; and (3) the number of plates in each component of grilles frame is consistent with the global design of phase I. The improved structural performance analysis is provided as follows:

PBL Shear Connectors

Figure 15 shows the Mises stress distribution of the PBL shear connectors. The Mises stress of the PBL shear connectors is less than that of the yield strength, and the maximum stress is 28 MPa, which occurs at the 12th row of the first column. For the PBL shear connectors in the same row, the stress on the side close to the wallboard is greater than that on the side far from the wallboard, and shear of the first row is the most disadvantageous.

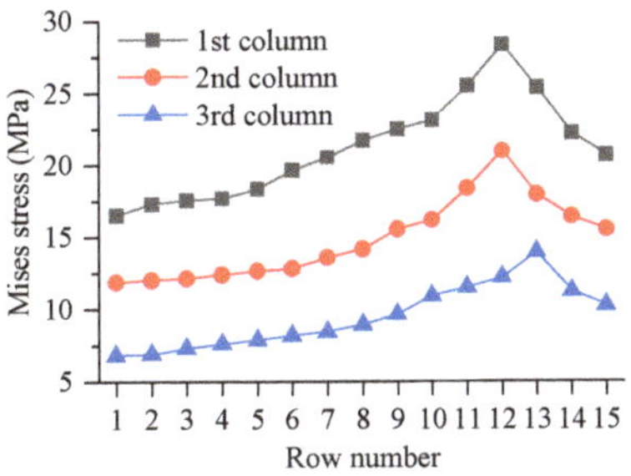

Figure 15. Mises stress distribution of the PBL shear connectors (unit: Pa).

Figure 16 shows the shear force and slip of PBL shear connectors at the first column. The shear force of PBL connectors is first increased, and then decreased with the increase in the row number. The maximum slipping value is 0.05 mm, thereby showing a gradually decreasing trend. The slip value is less than 0.2 mm, thereby meeting the design requirements of Formula (6).

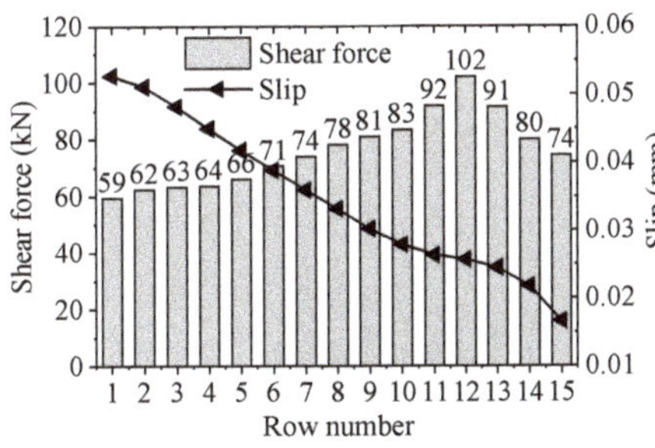

Figure 16. Shear force and slip of PBL shear connectors at 1st column.

U-Shaped Anchor Rebar

Three sets of anchor rebars are found on the inner side of the wallboard, where each set has six rebars for a total of 18 U-shaped anchor rebars. The stress value and group-sharing proportion of U-shaped anchor rebar are shown in Figure 17a,b.

As shown in Figure 17a, the maximum tensile stress of U-shaped anchor rebar is 228 MPa, which is less than the threshold value of 330 MPa. The maximum tensile stress appears at the first group of U-shaped anchor rebar near the loading side. The tensile stress gradually decreases because the U-shaped anchor rebar position is far from the loading side. The group-sharing proportion of U-shaped anchor rebar is uniform, indicating that all U-shaped anchor rebars undertake the out-of-plane deformation of the wallboard.

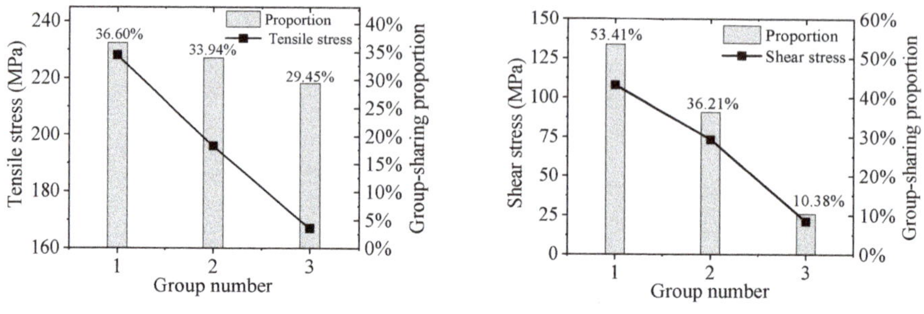

(a) (b)

Figure 17. Stress and group-sharing proportion of U-shaped anchor rebar: (**a**) Tensile stress and (**b**) shear stress.

As indicated in Figure 17b, the maximum shear stress at the weld of U-shaped anchor rebar is 107 MPa, which is less than the limit value of 175 MPa and meets the safety requirements. The maximum shear stress at the weld appears at the first group of U-shaped anchor rebar near the loading side. The maximum shear stress significantly decreases with the U-shaped anchor rebar position away from the loading side. At the same time, the group-sharing proportion among the groups is uneven. The group-sharing proportion among the first group is more than 50%, and the proportion of the third group is only 10%, thereby indicating that the U-shaped anchor rebars provide shear bearing reserve.

Grille Frame

Figure 18 shows the maximum Mises stress change in the grid frame during construction. The Mises stress of the grille frame under thrust force is less than the yield strength of 345 MPa. The maximum Mises stress is 201 MPa, which appears at the weld between the transverse and longitudinal plates near the loading side. This condition is mainly because the longitudinal plate is bent and deformed, resulting in the stress concentration caused by the limited internal displacement. Meanwhile, the Mises stress in the rest of the grille frame is significantly reduced. The plates of the entire grille frame have low stress, thereby meeting the criterion in Table 2. Meanwhile, the maximum displacement of the grille frame in the Y direction is 0.365 mm, which is less than the limit value calculated using Formula (7).

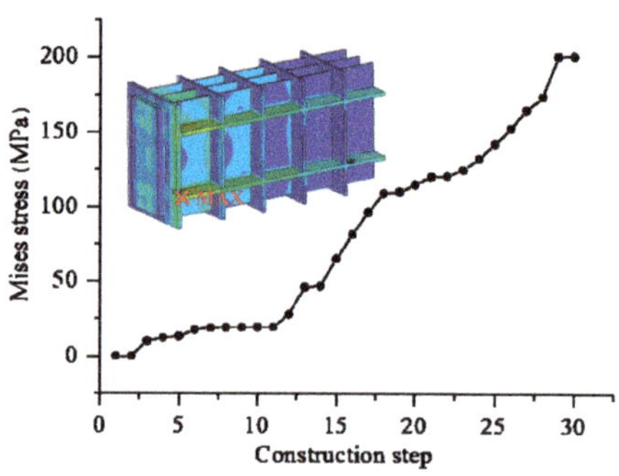

Figure 18. Mises stress of the grille frame.

Concrete Anchorage Zone

As shown in Figure 19, the maximum principal tensile and compressive stresses in the concrete anchorage zone are 1.46 and 17.4 MPa, which are less than the limit values of 1.86 and 22.4 MPa, respectively. The maximum tensile and compressive stresses occur at the joint of the wallboard and concrete, where an obvious stress concentration phenomenon occurs, and the tensile and compressive stresses of concrete far from this area significantly decrease.

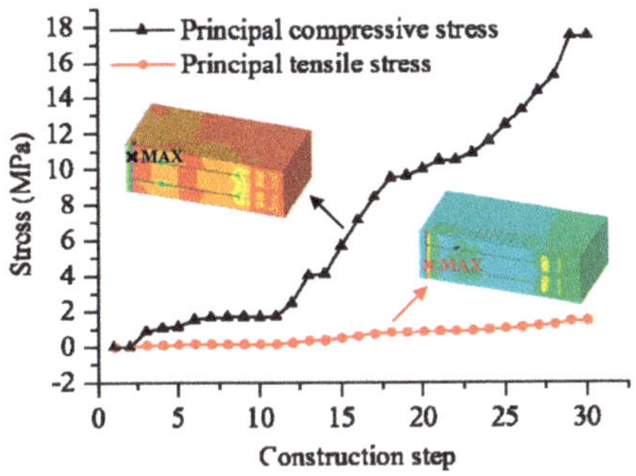

Figure 19. Stress in anchorage zone of concrete side span beam.

4.4. Field Test and Verification

As shown in Figure 20a,b, the designed girder–pylon thrust shoulders have been successfully applied to help facilitate the implementation of sustainable construction of the Dongtiao River Bridge. For field tests, three stress measuring points are arranged on the grille frame to monitor the stress changes during the construction. Figure 20c,d shows the comparison of the measured and calculated stress values at points 1, 2, and 3. The grille frame stress is constantly in the safe range, and the variation tendencies of the measured and calculated values are the same. The measured values are slightly larger than the calculated values because of the measurement errors and construction load uncertainty. The errors of the calculated and measured values are within 10 MPa and are in good agreement, verifying the accuracy of the proposed analytical framework.

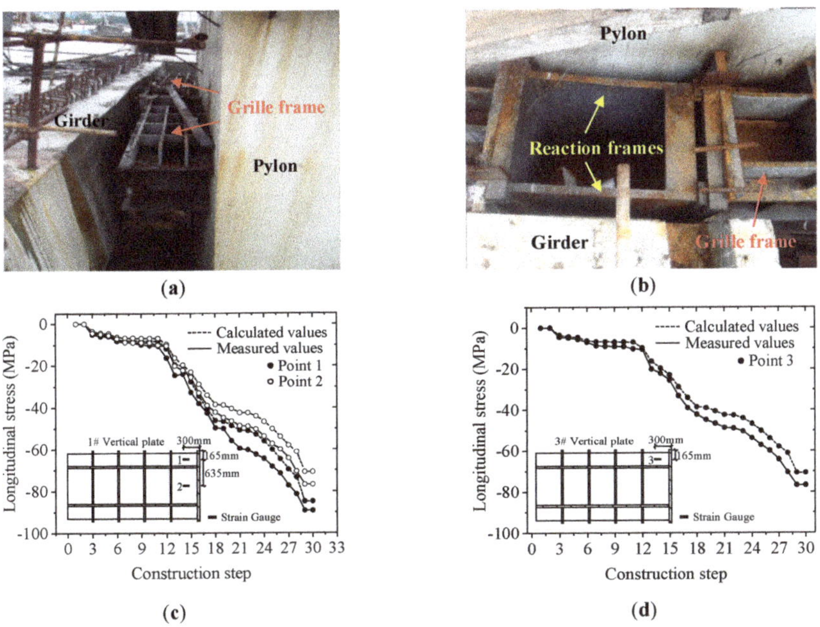

Figure 20. Field tests: (**a**) Girder-pylon antithrust system; (**b**) force transmission brace; and (**c**,**d**) calculated and measured results of the longitudinal stresses of the grille frame.

5. Result Discussion on Construction Sustainability

The proposed TPA technology enables a construction sequence that is similar to the construction of conventional suspension bridge, where the main cable is first erected, and the mid-span girder is lifted in sections and connected to the hangers (Figure 21a). In this way, the ship navigation during the construction period avoids interruptions (Figure 21b). Without the need for temporary supports, the TPA technology enables to minimize environmental impact and traffic disruption during the construction stage.

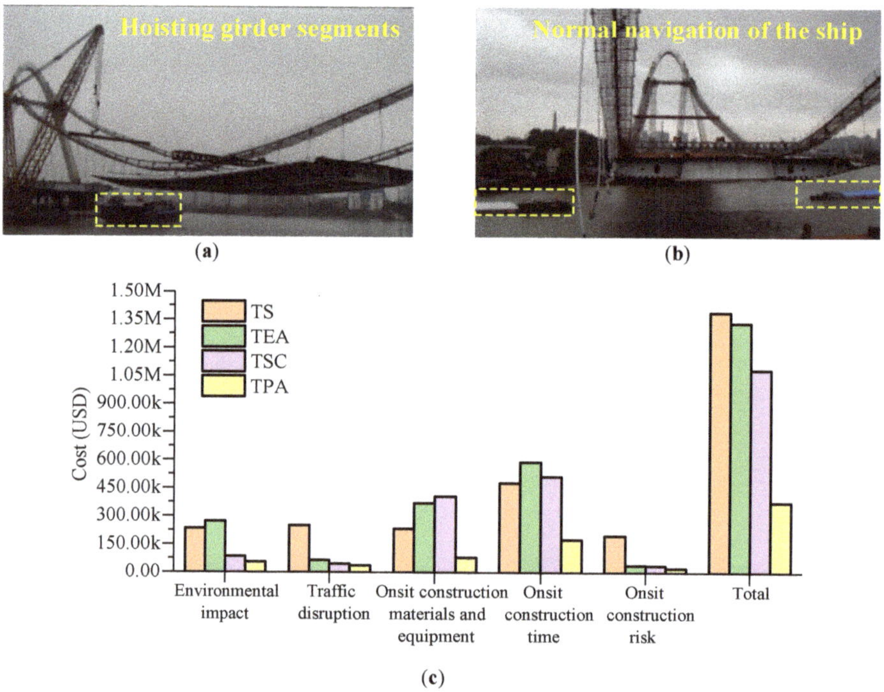

Figure 21. Construction sustainability evaluation: (**a**) Hoisting girder segments; (**b**) channel operation during construction; and (**c**) economic evaluation for construction.

A comprehensive onsite construction cost is evaluated for the engineering practice. As shown in Table 4, the traffic disruption has been minimized to 76 hours during the entire 4 months period of girder erection. The environmental impact on the navigation channel can be eliminated. The total onsite construction time is significantly reduced to 21 days. In Table 4, materials for the GPAS include steel (ton) and reinforced bar (ton), and the equipment for erection and removal of GPAS includes arc welder (set) and plasma cutting machine (set).

To compare sustainability among four construction technologies, the comprehensive onsite construction costs are calculated. As shown in Figure 21c, the total cost of TPA technology is the least, which accounts for 27%, 28%, and 34% of the total cost of TS, TEA, and TSC technology, respectively. In particular, the TPA technology significantly reduces the cost of environmental and traffic impact compared with other technologies, indicating that the impact on the environment and traffic is relatively small. Meanwhile, in terms of onsite materials and equipment, construction time and risk, TPA technology also offers obvious economic advantages.

Table 4. Comprehensive onsite construction cost of TPA technology.

Division of Construction			Unit	Quantity	Unit price (USD)	Cost (USD)	Sum (USD)
Environmental impact (C_1)						56,838	56,838
Traffic disruption (C_2)			Hour	76	412	31,312	31,312
Onsite construction materials and equipment (C_3)	GPAS Materials	Steel	ton	7.87	584	4596	80,400
		Reinforcing bars	ton	0.8	541	433	
	Equipment	Arc welder	Set	2	5870	11,740	
		Plasma cutting machine	Set	2	21,428	42,856	
		Other	Set	15	1385	20,775	
Onsite construction time (C_4)			Day	21	8405	176,505	176,505
Onsite construction risk (C_5) TPGA						22,600	22,600
Total (C)							367,655

6. Conclusions

To promote sustainable construction within real-world application, a comprehensive research is studied in the paper on sustainable construction by using novel technology for self-anchored suspension bridge. The main conclusions and innovation points are summarized as follows:

1. A novel temporary pylon-anchor (TPA) technology is proposed to minimize environmental impact and traffic disruption. A novel girder-pylon antithrust system (GPAS) is developed to achieve the engineering application for the TPA technology. The new TPA and GPAS technology could efficiently improve the construction sustainability.

2. A practical assessment method and index of sustainability on bridge construction technology are established to facilitate decision making for sustainable construction scheme. This paper creates an evaluation index system by considering different aspects: Environmental impact, traffic disruption, onsite construction materials and equipment, time, and risk.

3. A two-phase analytical approach for the GPAS is proposed by considering the global design and parametric study. In phase I, the global design of thrust shoulder is performed using a set of specially derived practical formulas. Various performance indicators are established to ensure the applicability and global reliability of the thrust shoulder. In phase II, the local stress distribution is improved based on effects of different parameters induced by grouped parametric analyses using 3D elaborate finite element analysis, which simulates the nonlinear surface contact behavior to consider the friction effect between the steel and the concrete.

4. The applicability and rationality of the proposed novel construction technology are illustrated by the successful application in real-world engineering. The field tests and sustainability assessment show that the proposed sustainability assessment method and analytical approach can facilitate the implementation of sustainable construction for self-anchored suspension bridge.

All in all, this research lays a solid and comprehensive basis to promote construction sustainability for self-anchored suspension bridge.

Author Contributions: X.W. (Xiaoming Wang) conceived and wrote the paper; X.W. (Xudong Wang) wrote the paper; Y.D. edited and improve the paper; C.W. provided the project data. All authors have read and agreed to the published version of the manuscript.

Funding: This research was funded by the Natural Science Foundation of Shannxi Province, grant number 2020JM-219, the China Postdoctoral Science Foundation, grant number 2019M653519, and the National Key Research and Development Project of China, grant number 2018YFB1600300.

Conflicts of Interest: The authors declare no conflict of interest.

References

1. Penadés-Plà, V.; Martí, J.V.; García-Segura, T.; Yepes, V. Life-Cycle Assessment: A Comparison between Two Optimal Post-Tensioned Concrete Box-Girder Road Bridges. *Sustainability* **2017**, *9*, 1864. [CrossRef]

2. Frangopol, D.M.; Dong, Y.; Sabatino, S. Bridge life-cycle performance and cost: Analysis, prediction, optimisation and decision-making. *Struct. Infrastruct. Eng.* **2017**, *13*, 1239–1257. [CrossRef]
3. Kripka, M.; Yepes, V.; Milani, C. Selection of Sustainable Short-Span Bridge Design in Brazil. *Sustainability* **2019**, *11*, 1307. [CrossRef]
4. Khan, M.A. *Accelerated Bridge Construction: Best Practices and Techniques*; Butterworth-Heinemann: Oxford, UK, 2014.
5. Dong, Y. Performance assessment and design of ultra-high performance concrete (UHPC) structures incorporating life-cycle cost and environmental impacts. *Constr. Build. Mater.* **2018**, *167*, 414–425. [CrossRef]
6. Gilchrist, A.; Allouche, E.N. Quantification of social costs associated with construction projects: State-of-the-art review. *Tunn. Undergr. Space Technol.* **2005**, *20*, 89–104. [CrossRef]
7. Grierson, D. Towards a sustainable built environment. *CIC Start Online Innov. Rev.* **2009**, *1*, 70–77.
8. Wang, Z.; Jin, W.-L.; Dong, Y.; Frangopol, D.M. Hierarchical life-cycle design of reinforced concrete structures incorporating durability, economic efficiency and green objectives. *Eng. Struct.* **2018**, *157*, 119–131. [CrossRef]
9. Goh, C.S.; Chong, H.-Y.; Jack, L.; Faris, A.F.M. Revisiting triple bottom line within the context of sustainable construction: A systematic review. *J. Clean. Prod.* **2020**, *252*, 119884. [CrossRef]
10. Anwar, G.A.; Dong, Y.; Zhai, C. Performance-based probabilistic framework for seismic risk, resilience, and sustainability assessment of reinforced concrete structures. *Adv. Struct. Eng.* **2019**. [CrossRef]
11. Gou, H.; Wang, W.; Shi, X.; Pu, Q.; Kang, R. Behavior of steel-concrete composite cable anchorage system. *Steel Compos. Struct.* **2018**, *26*, 115–123.
12. Wang, X.; Frangopol, D.M.; Dong, Y.; Lei, X.; Zhang, Y. Novel Technique for Configuration Transformation of 3D Curved Cables of Suspension Bridges: Application to the Dongtiao River Bridge. *J. Perform. Constr. Facil.* **2018**, *32*, 04018045. [CrossRef]
13. Gil, H.; Choi, Y. Cable Erection Test at Splay Band for Spatial Suspension Bridge. *J. Bridg. Eng.* **2002**, *7*, 300–307. [CrossRef]
14. Sun, J.; Manzanarez, R.; Nader, M. Suspension Cable Design of the New San Francisco–Oakland Bay Bridge. *J. Bridg. Eng.* **2004**, *9*, 101–106. [CrossRef]
15. Nie, J.; Zhou, M.; Wang, Y.-H.; Fan, J.-S.; Tao, M.-X. Cable Anchorage System Modeling Methods for Self-Anchored Suspension Bridges with Steel Box Girders. *J. Bridg. Eng.* **2014**, *19*, 172–185. [CrossRef]
16. Nader, M. Accelerated Bridge Construction of the New Samuel De Champlain Bridge. *J. Bridg. Eng.* **2020**, *25*, 05019015. [CrossRef]
17. Siwowski, T.; Wysocki, A. Horizontal Rotation via Floatation as an Accelerated Bridge Construction for Long-Span Footbridge Erection: Case Study. *J. Bridg. Eng.* **2015**, *20*, 05014014. [CrossRef]
18. Sun, J.F.; Fang, J.H.; Tan, Q.Q.; F.Li, L. Introduction of novel construction method for superstructure of self-anchored suspension bridge. *Prestress Technol.* **2004**, 25–29.
19. Ultimate capacity of narrow type steel box section for railway self-anchored suspension bridge under bias compression. In *Ultimate Capacity of Narrow Type Steel Box Section for Railway Self-Anchored Suspension Bridge under Bias Compression*; Hong Kong Institute of Steel Construction: Hong Kong, China, 2019; Volume 15, pp. 173–184.
20. Verkehr, B.F.; Wirtschaft, M.F.; Rheinland, L. *Denkschrift zur Verkehrsübergabe der Wiederhergestellten Autobahnbrücke über den Rhein in Rodenkirchen bei Köln am 9.Dezember 1954*; Springer Science and Business Media LLC: Berlin, Germany, 1954.
21. Van Goolen, D. *Self-Anchored Suspension Bridges*; Delft University of Technology: Delft, The Netherlands, 2006.
22. Navarro, I.J.; Yepes, V.; Martí, J.V. Life Cycle Cost Assessment of Preventive Strategies Applied to Prestressed Concrete Bridges Exposed to Chlorides. *Sustainability* **2018**, *10*, 845. [CrossRef]
23. Penadés-Plà, V.; García-Segura, T.; Martí, J.V.; Yepes, V. An Optimization-LCA of a Prestressed Concrete Precast Bridge. *Sustainability* **2018**, *10*, 685. [CrossRef]
24. Chang, Y.; Yang, Y.; Dong, S. Comprehensive Sustainability Evaluation of High-Speed Railway (HSR) Construction Projects Based on Unascertained Measure and Analytic Hierarchy Process. *Sustainability* **2018**, *10*, 408. [CrossRef]
25. Seo, S.-Y.; Lee, B.; Won, J. Comparative Analysis of Economic Impacts of Sustainable Vertical Extension Methods for Existing Underground Spaces. *Sustainability* **2020**, *12*, 975. [CrossRef]

26. Nie, J.; Tao, M.; Fan, J. Research on Cable Anchorage Systems for Self-Anchored Suspension Bridges with Steel Box Girders. *J. Bridg. Eng.* **2011**, *16*, 633–643. [CrossRef]
27. Deng, W.; Xiong, Y.; Liu, D.; Zhang, J. Static and fatigue behavior of shear connectors for a steel-concrete composite girder. *J. Constr. Steel Res.* **2019**, *159*, 134–146. [CrossRef]
28. Leroy, A.L.; Peter, G. Mechanics of bond and slip of deformed bars in concrete. *J. Proc.* **1967**, *64*, 711–721.
29. Zhao, C.; Li, Z.; Deng, K.; Wang, W. Experimental investigation on the bearing mechanism of Perfobond rib shear connectors. *Eng. Struct.* **2018**, *159*, 172–184. [CrossRef]
30. Huang, Q. *Design Principle of Bridge Steel-Concrete Composite Structure*; China Communications Press Co., Ltd.: Beijing, China, 2017.
31. MOT. *GB 50917-2013. Code for Design of Steel and Concrete Composite Bridges*; CABP: Beijing, China, 2013.
32. Lou, X.; Wu, H.; Cui, B. *Designing and Construction of Suspension Bridge Main Cable Anchor Transferring Load Distributively*; China Communications Press Co., Ltd.: Beijing, China, 2012.
33. MOT. *JTG D62-2004 Specifications for Design of Highway Reinforced Concreteand Prestressed Concrete Bridges and Culverts*; CCP: Beijing, China, 2004.
34. MOT. *GB 50017-2017 Standard for Design of Steel Structures*; CABP: Beijing, China, 2017.
35. MOT. *JTG/T F50-2011 Technical Specification for Construction of Highway Bridge and Culverts*; CCP: Beijing, China, 2011.
36. Ahn, J.-H.; Lee, C.-G.; Won, J.-H.; Kim, S.-H. Shear resistance of the perfobond-rib shear connector depending on concrete strength and rib arrangement. *J. Constr. Steel Res.* **2010**, *66*, 1295–1307. [CrossRef]
37. Wang, S.; He, J.; Liu, Y.; Li, C.; Xin, H. Shear capacity of a novel joint between corrugated steel web and concrete lower slab. *Constr. Build. Mater.* **2018**, *163*, 360–375. [CrossRef]
38. Gu, J.-C.; Liu, D.; Deng, W.-Q.; Zhang, J. Experimental study on the shear resistance of a comb-type perfobond rib shear connector. *J. Constr. Steel Res.* **2019**, *158*, 279–289. [CrossRef]
39. Li, Z.; Zhao, C.; Shu, Y.; Deng, K.; Cui, B.; Su, Y. Full-scale test and simulation of a PBL anchorage system for suspension bridges. *Struct. Infrastruct. Eng.* **2019**, *16*, 452–464. [CrossRef]
40. Willam, K.J. Constitutive model for the triaxial behaviour of concrete. *Proc. Intl. Assoc. Bridge Structl. Eng.* **1975**, *19*, 1–30.
41. He, J.; Liu, Y.; Lin, Z.; Chen, A.; Yoda, T. Shear behavior of partially encased composite I-girder with corrugated steel web: Numerical study. *J. Constr. Steel Res.* **2012**, *79*, 166–182. [CrossRef]

sustainability

MDPI

Article

Black Hole Algorithm for Sustainable Design of Counterfort Retaining Walls

Víctor Yepes [1,*,†], José V. Martí [1,†] and José García [2,†]

1 Institute of Concrete Science and Technology (ICITECH), Universitat Politècnica de València, 46022 València, Spain; jvmartia@cst.upv.es
2 Pontificia Universidad Católica de Valparaíso, 2362807 Valparaíso, Chile; jose.garcia@pucv.cl
* Correspondence: vyepesp@cst.upv.es
† These authors contributed equally to this work.

Received: 25 February 2020; Accepted: 27 March 2020; Published: 1 April 2020

Abstract: The optimization of the cost and CO_2 emissions in earth-retaining walls is of relevance, since these structures are often used in civil engineering. The optimization of costs is essential for the competitiveness of the construction company, and the optimization of emissions is relevant in the environmental impact of construction. To address the optimization, black hole metaheuristics were used, along with a discretization mechanism based on min–max normalization. The stability of the algorithm was evaluated with respect to the solutions obtained; the steel and concrete values obtained in both optimizations were analyzed. Additionally, the geometric variables of the structure were compared. Finally, the results obtained were compared with another algorithm that solved the problem. The results show that there is a trade-off between the use of steel and concrete. The solutions that minimize CO_2 emissions prefer the use of concrete instead of those that optimize the cost. On the other hand, when comparing the geometric variables, it is seen that most remain similar in both optimizations except for the distance between buttresses. When comparing with another algorithm, the results show a good performance in optimization using the black hole algorithm.

Keywords: CO_2 emission; earth-retaining walls; optimization; black hole; min–max discretization

1. Introduction

The economic sustainability and social development of most countries depend directly on the reliable and lasting behavior of their infrastructure [1]. Infrastructures have special relevance because they strongly influence economic activity, growth, and employment. However, these activities have a significant impact on the environment, have irreversible effects, and can compromise the present and future of society. It is known that construction is a carbon-intensive industry [2]; therefore, the minimization of emissions is increasingly important to reduce the environmental impact of construction projects. Looking further, we find that cement is one of the materials that generates large amounts of emissions. A large amount of emissions is mainly due to the calcination process of the limestone together with the high energy demand necessary for its production [3]. Therefore, the optimization of structures that use large amounts of cement is critical. Consequently, it is essential to develop lines of research in sustainable construction [4,5], energy consumption [6,7], and in the study of the cycle of the CO_2 emissions made by concrete structures [8]. The great challenge is to have an infrastructure capable of maximizing its social benefit without compromising its sustainability [9].

The standard procedure [10] to perform the economic estimation of structures is based on a trial-and-error approach. This approach is necessary because the dimensions of the cross-section and the material grades are defined a priori. This involves analyzing the stresses and calculating the effort required in each material configuration. This standard procedure is slow and expensive.

Particularly in the case of a wall, there are a number of parameters that must be taken into account, such as boundary conditions, the type of filling, the friction base angles, load capacity of the base of the ground, and surcharge loads.

On the other hand, the emissions of CO_2 constitute an important portion of the contribution to the global warming process of the planet. In particular, approximately 40%–50% of global GHG emissions are generated by construction [11]. In addition, the cement industry releases 5% of global GHG emissions [12]. Therefore, sustainability and specifically the use of high-carbon-footprint products in structural engineering is a relevant line of research. At present, the consumption of CO_2 has been investigated by considering it as an objective function within an optimization problem. Particularly in [13], the optimization of CO_2 consumption in walls was analyzed using the harmony search algorithm. In [14], the impact of CO_2 on cantilever retaining walls was studied. A CO_2 and cost analysis in precast–prestressed concrete road bridges was developed in [11]. The study of a sustainable prototype for reinforced concrete columns was proposed in [15]. In this study, heuristic methods of optimization of CO_2 emissions were applied. In [16], the importance of the criteria that define social sustainability was analyzed. These criteria considered the complete life of infrastructure. The social sustainability of infrastructure projects was tackled in [17] through the use of Bayesian methods. In [18], an optimization was studied based on the cost criterion of reinforced concrete-retaining walls. The study considered using different types of load capacity estimates. The life cycle assessment of earth-retaining walls was analyzed in [19,20]. In [21], a stochastic analysis of the emissions of CO_2 was developed and applied to construction sites. The relevance of carbon emissions and low-emission window films on energy spending was studied in [22] and applied to an existing UK hotel building. In [23], a methodology was built for the investigation of carbon emissions in a hospital building, which allowed the generation of an information model and an evaluation of its life cycle. In [24], an optimization of reinforced concrete columns was optimized considering several environmental impact assessment parameters.

To solve large-scale and highly non-linear optimization problems, metaheuristic algorithms, which have demonstrated their efficiency and versatility, have been used [25]. There has been a lot of research in the area of metaheuristics in recent years, and a large number of algorithms have been developed.

Most of these algorithms have been inspired by physical, chemical, and biological systems [26]. As examples of these heuristic search algorithms that belong to this category, we find particle swarm optimization (PSO), harmony search (HS) [27], threshold acceptance (TA), simulated annealing (SA) [28,29], threshold acceptance (TA), genetic algorithms (GA), ant colonies (ACO), genetic algorithms (GA), whale optimization (WO), cuckoo search (CS), and black hole (BH), among others. The use of metaheuristic algorithms in sustainability and construction has been used, for example, in [30], where they applied heuristic methods to the study of reinforced concrete road vaults; in addition, a metaheuristic was applied to the problem of decision making in [31]. In [32], a mathematical model was developed with the goal of improving environmental economic policies. This model was applied to a protection zone. A survey of the social sustainability evaluation applied to the infrastructures was developed in [33]. However, we must consider that many of the metaheuristics work in continuous spaces. Researchers have been developing techniques to obtain discrete or binary versions of these metaheuristics. For more details on binarization techniques, the reader can consult [34–36].

In this article, inspired by the aforementioned, we explore the application of a binary black hole algorithm to solve the optimization of counterfort retaining walls. The contributions of this work are detailed below:

1. An adaptation of the BH optimization technique to work in discrete environments. Naturally, BH works in continuous spaces. Here, an adaptation is proposed using the concept of optimal approach velocity and a min–max normalization, which allows the velocity to be transformed into a transition probability.

2. The application of the discrete version of BH to the counterfort retaining walls optimization problem. This optimization considers the objective function, the costs, and the CO_2 separately.
3. The impact of the relevant design variables is studied, both in costs and in CO_2 emission.

In Section 2, the optimization problem, the variables involved, and the restrictions are defined. The algorithm that executes the optimization is detailed in Section 3. The experiments and results obtained are shown in Section 4. Finally, in Section 5, the conclusions and new lines of research are summarized.

2. Problem Definition

In this section, we will detail the buttressed earth-retaining walls optimization problem. In Section 2.1, we will give the definition of the optimization problem. Later, in Section 2.2, we will detail the design variables. Then, the design parameters will be described in Section 2.3, and finally, in Section 2.4, we define the constraints.

2.1. Optimization Problem

The optimization problem considers two objective functions, which will be addressed independently. The first function corresponds to the cost (p_i) of the wall construction, expressed in euros. The second one considers the CO_2 equivalent emissions units (e_i). These construction units correspond to formwork, materials, earth-fill, and excavation. The calculations of emissions and cost are based on a 1 m wide strip. The emission and cost unit values were obtained from [13,37] and are shown in Table 1. Then, in a general way, our optimization problem is defined according to the Equation (1).

$$O(x) = \sum_{i=1}^{r} a_i x_i, \tag{1}$$

where x represents the measurement of the set of decision variables, and $a \in \{c, e\}$ corresponds to cost or emissions. Additionally, the optimization problem is subject to a set of restrictions determined by the service (SLS) and ultimate (ULS) state limits.

Definition 1 ([10]). *Definition of costs and emissions.*

Table 1. Unit breakdown by unit cost, emissions, and cost. Source: [10].

Unit	Cost (€)	Emissions (CO_2-eq)
kg steel B400	0.56	3.02
kg steel B500	0.58	2.82
m^3 of concrete in stem		
C25/30	56.66	224.34
C30/37	60.80	224.94
C35/45	65.32	265.28
C40/50	70.41	265.28
C45/55	75.22	265.91
C50/60	80.03	265.95
m^2 stem formwork	21.61	1.92
m^3 of backfill	5.56	28.79
m^3 of concrete in foundation		
C25/30	50.65	224.34
C30/37	54.79	224.94
C35/45	59.31	265.28
C40/50	64.40	265.28
C45/55	69.21	265.91
C50/60	74.02	265.95

2.2. Problem Design Variables

In the design of the buttressed retaining wall to be studied, three groups of variables are defined: The geometric variables, the concrete and steel grade, and the passive reinforcement of the footing and the stem of the wall. In total, there are 32 variables, where each of these has a discrete number of possibilities in the group of concrete and steel grades. For the realization of this study, Portland cement was considered, as it is the most widely applied in the construction of walls. Types C25/30 to C50/60 were considered in discrete 5 MPa intervals. Other types of cements with different additions could give different environmental results. In the case of steel, the B500S and B400S types were considered. In its production phase, the recycled scrap is considered approximately 40% of the total, and the manufacturing is done in an electric arc furnace. Therefore, the technology in the manufacture of steel and cement, as well as the nature of cement additions, influence CO_2 emissions [38] and affect the values in Table 1.

In the group of geometric variables, there is the stem thickness (b), the footing thickness (c), the distance between buttresses (d), the toe length (p), the buttress thickness (e_c), and heel length (t). The last group of variables correspond to the passive reinforcement of the footing and the stem of the wall. Figures 1 and 2 show the 24 variables that define the problem.

The steel bar diameters and the number of bars determine the reinforcement. A1, A2, and A3 define the three reinforcement flexural bars that contribute to the main bending of the stem. A4 defines the vertical reinforcement of the foundation at the rear side of the stem, up to a height L1. A5 provides the secondary longitudinal reinforcement for shrinkage and thermal effects on the stem. A6 provides longitudinal buttress reinforcement. A7 and A8 define the area of reinforcement bracket from the bottom of the buttress, vertically and horizontally, respectively. A9 provides the upper heel reinforcement. A10 defines the longitudinal effects on the toe. A11 provides the bottom heel reinforcement, and A12 defines the shear reinforcement on the footing. Table 2 shows the ranges of the values for the 32 variables in the problem, and the set of combinations of these values constitutes the solution space.

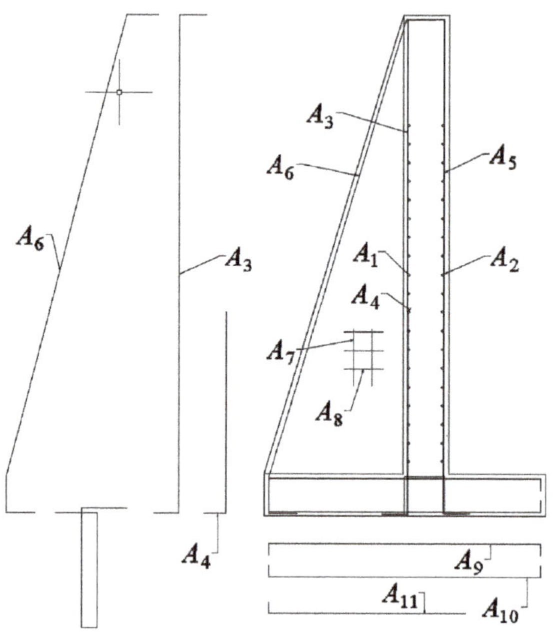

Figure 1. Set of reinforcement variables. Source: [10].

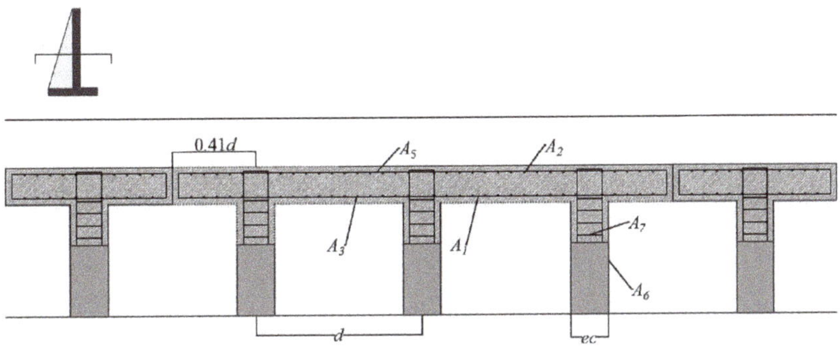

Figure 2. Buttressed wall. Cross-section of the floor. Source: [10].

Definition 2 ([10]). *Definition of design variables.*

Table 2. Design variables. Source: [10].

Variables	Lower Bound	Upper Bound	Increment	N of Values
c	H/20	H/5	5 cm	f(H) [1]
d	H/5 cm	2H/3	5 cm	f(H) [1]
b	25 cm	122.5	2.5 cm	40
p	20 cm	610	10 cm	60
t	20 cm	905	15 cm	60
e_c	25 cm	122.5	2.5 cm	40
f_{ck}	25, 20, 25, 40, 45, 50			7
f_{yk}	400, 500			2
A_1 to A_{10}	6, 8, 10, 12, 16, 20, 25, 32			8
	1 steel rebar	12 rebars	2 rebars	6
A_{11} to A_{12}	6, 8, 10, 12, 16, 20, 25, 32	8		
	1 steel rebar	4 rebars	10 rebars	7

[1] Depending on height.

2.3. Problem Design Parameters

The design parameters of the problem are the data that will take constant values in the optimization process. The main design parameters are shown in Figure 3. H and H2 represent the wall height and the soil depth in front of the wall. The base coefficient of friction is represented by μ, the maximum support pressure for operating conditions is σ, and the backfill slope at the top of the rod is represented by β. The maximum bearing capacity considered in the soil foundation is the ultimate bearing pressure divided by the bearing capacity safety factor. The pressure angle of the earth is determined by P (γ, ϕ, δ), corresponding to the density, the angle of friction, and the angle of internal friction. Finally, a part of ϕ defines the roughness between the wall and the filling. The values of the problem design parameters are shown in Table 3.

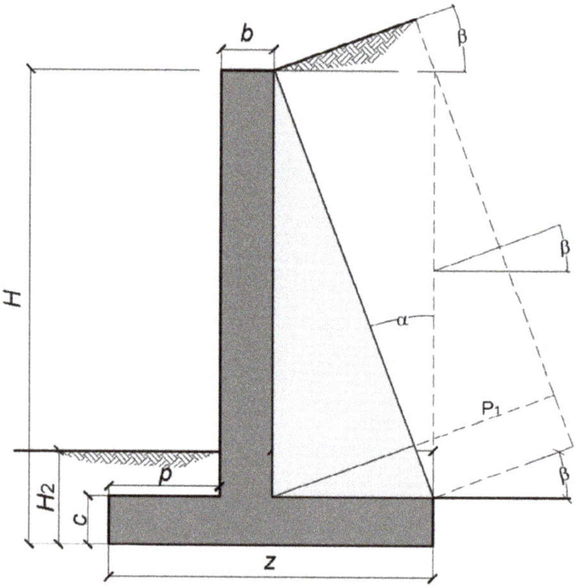

Figure 3. Problem design parameters. Source: [10].

Definition 3 ([10]). *Definition of design parameters.*

Table 3. Problem design parameter values.

Parameter Considered		Value
Bearing capacity		0.3 MPa
Fill slope		0
Foundation depth, H2		2 m
Uniform load on top of the fill, γ		10 kN/m^2
Base-friction coefficient, μ		tg 30°
Wall-fill friction angle, δ		0°
Safety coefficient:		
	against sliding, γ_{fs}	1.5
	against overturning, γ_{fo}	1.8
	for loading (EHE)	Normal
	of steel (ULS)	1.15
	of concrete (ULS)	1.5
Ambient exposure		IIa

2.4. Problem Constraints

The feasibility of the structure is verified in accordance with the Spanish technical standards defined in [39] and the recommendations detailed in [40]. The flexural and shear limit states are checked. The structure is verified in accordance with the approach specified in [41]. To check the structure limit states, a uniform surface load at the top of the fill is considered [42]. For the active earth pressure calculation, the surface loads and fill are considered. The major forces in wall analysis consider wall weight, heel backfill load, surface load, earth pressure, weight at the front toe, and passive resistance against the toe. Buttresses receive a load equivalent to the product of the distance between the buttresses by the pressure distribution in the stem.

The structural model considers that the upper part of the stem works as a cantilever, while the lower part of the stem is strongly coerced by two elements: The footing and the lower part of the buttress, located at the rear of the stem. Calculations of the bending moments are taken in the midsection between the buttresses and are given by B_1 and B_2, described in Equations (2) and (3), respectively.

$$B_1 = -0.03p_1d(H-c) \tag{2}$$

$$B_2 = -0.0075p_1d(H-c). \tag{3}$$

B_1 is the bending moment at the connection of the stem to the footing, B_2 is the maximum bending moment on the stem, and p_1 represents the pressure over the slab on the upper side of the footing. When the spacing of the buttresses is less than 50% of the height, Equation (4) defines the shear resistance (s) at the connection of the plate to the footing. For an accurate estimate of the moments in each section of the stem, as a result of the vertical bending stress in the stem, the trapezoidal pressure distribution is considered [41]. A total of 50% of the maximum pressure at the top of the foundation is taken as the maximum value. Taking into account the vertical bending moment in the upper quarter part of the stem may be insignificant due to the involvement of [41].

$$s = 0.4p_1d \tag{4}$$

Verification of the bending stress in the T-shaped horizontal cross-section is made considering the effective width according to [43]. Mechanical bending and shear capacity are evaluated using the equations expressed in [42]. In this manual, the construction limit states of the EHE Structural Concrete Code are considered. The checks against sliding, overturning, and soil stresses are carried out taking into account the effect of the buttresses, and are given in Equations (5)–(7). In the overturning check, Equation (5) ensures that the favorable overturning moments are sufficiently greater than the unfavorable overturning moments. In Equation (6), B_{of} is the total favorable overturning moment. In Equation (7), B_{ou} is the unfavorable total overturning moment, and the overturning safety factor is γ_{to} and is taken as 1.8 for frequent events. Equation (8) represents the reaction of soil against sliding. As μ is the base-friction coefficient, N' corresponds to the total sum of the ground and wall weights located at the heel and toe, and E_p determines the passive resistance against the toe defined by the Equation (9).

$$B_{of} - \gamma_{to}B_{ou} \geq 0 \tag{5}$$

$$B_{of} = N'(\frac{B}{2} - e_p) - E_p(H_t - h') \tag{6}$$

$$B_{ou} = E_h * h_e - E_v(\frac{B}{2} - f) \tag{7}$$

$$R = N'\mu + E_p \tag{8}$$

$$E_p = \frac{1}{2\gamma(H_t^2 - (H_t - c))^2} \frac{(1 + sin(\phi))}{(1 - sin(\phi))} \tag{9}$$

3. The Discrete Black Hole Algorithm

Using the analogy with a black hole, in [44], the Black Hole optimization algorithm was proposed. A region of space that concentrates a large amount of mass—therefore, there is no way for an object to escape from it—is considered a black hole. This same concept is used in the black hole optimization algorithm. In each iteration of the algorithm, best candidate is considered, and this is assigned as a black hole. Making the analogy, the selected candidate begins to attract other solutions. A solution is attracted to the black hole when the solution is too close to it, and disappears forever. A new solution is generated in this case. Then, the movement of a solution in the search space is conditioned by the black hole according to the Equation (10).

$$x_i^d(t+1) = x_i^d(t) + r[x_{BH}^d - x_i^d(t)], \forall i \in \{1, ..., N\}, \tag{10}$$

where N is the total number of solutions. $x_i^d(t)$ and $x_i^d(t+1)$ define the i-th component obtained in the iterations t and $t+1$. Finally, x_{BH}^d represents the d-th component of the best solution obtained. The black hole pseudo-code is detailed in Algorithm 1.

$$R = \frac{f_{BH}}{\sum_{i=1}^{N} f_i} \tag{11}$$

Algorithm 1 Black Hole Algorithm

```
Initialize solutions
Select black hole x_BH
while Iteration < MaxIteration do
    for all X ∈ particle do
        apply objective movement operator
        apply objective function f_X
        if f_X < f_BH then
            Replace the black hole with the new best solution
        end if
        if R > d(X, BH) then
            Generate random new solution
        end if
    end for
end while
```

The Discrete Algorithm

BH works in a natural way in continuous spaces, and, consequently, it must be adjusted to work properly in discrete spaces. The first step is to consider the velocity of the solution, which is $x(t+1) - x(t)$ and to normalize the vector using a min–max procedure. Subsequently, whether each component x^d of the solution should be modified or not is evaluated. This is done through comparison with a random number r_1. In the case that change is selected, the movement can increase the value (+1) or decrease it (−1). Finally, the selected value is compared with the value obtained by BH and remains with the minimum of the two. The pseudocode of the binary procedure is shown in Algorithm 2.

Algorithm 2 Discrete Algorithm

```
movement = 0
nx <–normalize solution
if r_1 < nx^d then
    if r_2 > 0.5 then
        movement = 1
    else
        movement = -1
    end if
    x^d = max(1,min(x^d_bh,x^d+movement))
end if
```

4. Results and Discussion

4.1. Wall Height Analysis

In this section, we will analyze the influence of wall height on the CO_2 emissions and cost of the structure. The experiment consists of varying the height of the wall from 6 to 15 m with intervals of 1 m, keeping the rest of the parameters constant. For each height, the algorithm was executed 30 times, minimizing the cost. Subsequently, the same procedure was performed, but minimizing CO_2 emissions. The results are shown in Figures 4 and 5 as well as Table 4.

In Figure 4, the green color represents the best value of the 30 executions that minimized the cost. The best value obtained by minimizing CO_2 emissions is shown in blue. The size of the point quantifies the deviation of the variable not optimized with respect to the optimum obtained. For example, for the height of 6 m, the optimum cost value was 593, and the value of CO_2 emissions obtained for that solution was 1270. The minimum emissions obtained were 1259; therefore, the emission deviation gets **Emission from optimum%** $= 100 * \frac{1270-1259}{1259}$.

The figure shows that for small heights, the deviations of CO_2 emissions from the optimum are minimal. From the height of 8 m, these begin to grow, obtaining the maximum deviation at the height of 11 m. When analyzing the deviations of the cost variable from its minimum, we see that it has a similar behavior to that in the case of CO_2 emissions.

In the second experiment, instead of plotting the minimum of the 30 executions, we will graph the average of these. The result is shown in Figure 5. The objective of this experiment is to analyze the stability of the solutions obtained by the algorithm when we consider different heights. In both series of points up to the height of 11 m, the average values vary less than 2% with respect to their minimum. From 11 m onwards, this variation increases, reaching over 10% with respect to the minimum in the case of 15 m. When we analyze the area of the points, we see that the series of points in green grows significantly unlike the blue one, which maintains a behavior similar to that of the Figure 4.

Table 4. Analysis of cost and emission optimization performed by the black hole (BH).

Height (m)	Opt. Cost	Emissions	Cost	Opt. Emissions	Cost from Optimum	Emissions from Optimum
6	593	1270	607	1259	2.36%	0.87%
7	679	1455	694	1441	2.21%	0.97%
8	774	1706	795	1659	2.71%	2.83%
9	912	2046	930	1998	1.97%	2.40%
10	1096	2589	1135	2478	3.56%	4.48%
11	1309	3249	1407	3060	7.49%	6.18%
12	1533	3841	1600	3716	4.37%	3.36%
13	1783	4602	1876	4471	5.22%	2.93%
14	2052	5523	2163	5291	5.41%	4.38%
15	2363	7116	2491	6788	5.42%	4.83%

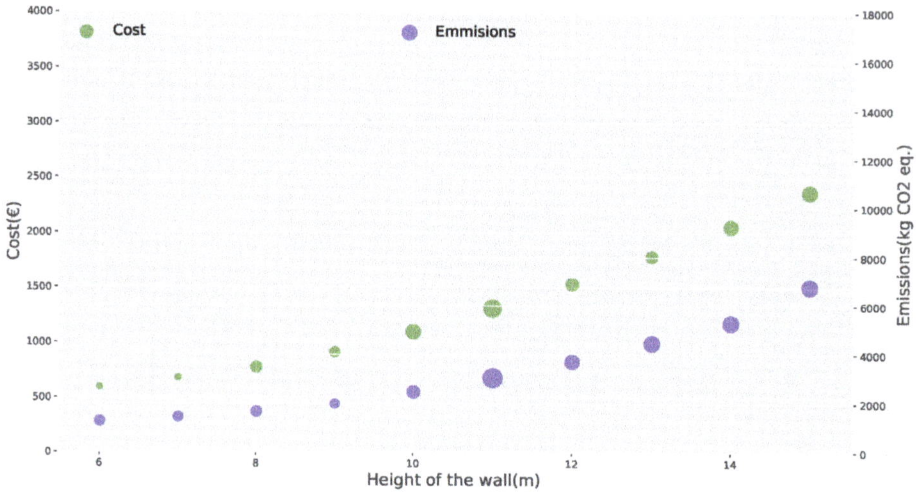

Figure 4. Variation and dispersion of best costs and CO_2 emissions by the height of the wall.

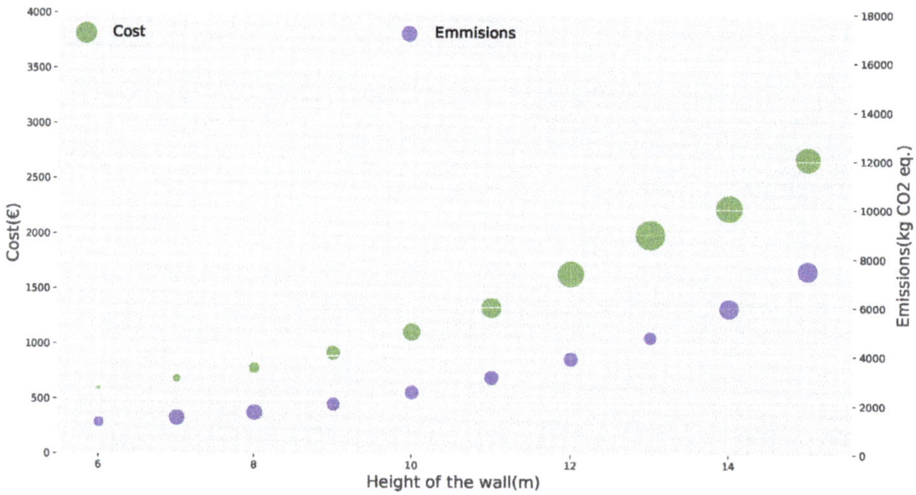

Figure 5. Variation and dispersion of average costs and CO_2 emissions by the height of the wall.

4.2. Design Variable Analysis

This section aims to study the variations of geometric, concrete, and steel variables when we perform cost and emission optimizations. In the first analysis, we will consider the geometric variables. The results are shown in the Figures 6 and 7. In Figure 6, the optimal values of the geometric variables for different heights of the wall are shown. The figure shows that stem, footing, and buttress thicknesses suffer zero or small variations as the height of the wall increases. The toe and heel lengths increase as the height of the wall increases and the distance between buttresses oscillates. In Figure 7, we compare the optimal geometric variables when performing the optimization by emissions and cost. For this analysis, we define a geometric ratio that corresponds to the quotient between the value obtained in the emissions optimization and the value obtained in the cost optimization. The figure shows that most of the values are the same in both optimizations, except for the variable of distance between buttresses. For this variable, the value obtained from emission optimization is always lower than that obtained from cost optimization. Additionally, we observe that for the height of 15 m, the variables of toe and heel length are modified in an inverse way.

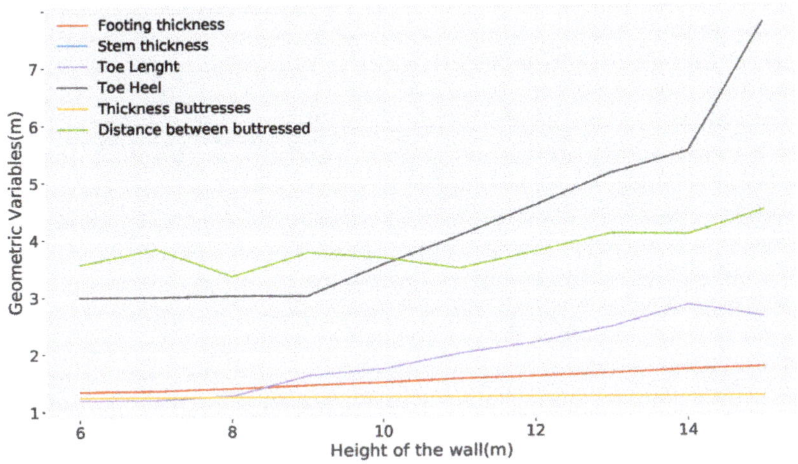

Figure 6. Geometric variables depending on the height of the wall.

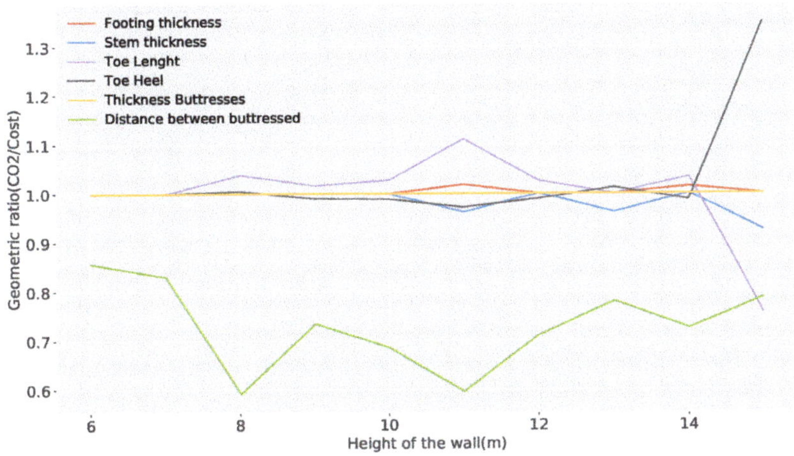

Figure 7. Ratio of geometrical variables.

In the second analysis, the variables of concrete and steel are studied. Each of these variables are separated in the material used in the stem and in the slab base. Then, the CO_2 emissions obtained from the optimization are calculated. The above is shown in Figure 8. The figure shows that for both concrete and steel, CO_2 emissions increase as the height of the wall increases. However, the interesting thing is that this increase is not linear, as the height increases the slope of the emissions increases as well.

Figure 9 compares the optimal concrete and steel variables when performing the optimization by emissions and cost. In both optimizations and for the different wall heights, the concrete and the resulting steel in all cases were the C25/30 and BS500. In this analysis, we define a concrete volume ratio that corresponds to the quotient between the value obtained in the emissions optimization and the value obtained in the cost optimization. The figure shows that the emissions of concrete in the stem are higher in the case of emissions optimization than in cost optimization. However, this increase is offset by a decrease in emissions in steel, both in the stem and in the slab base. In Table 5, the values of concrete and steel are attached for both optimizations.

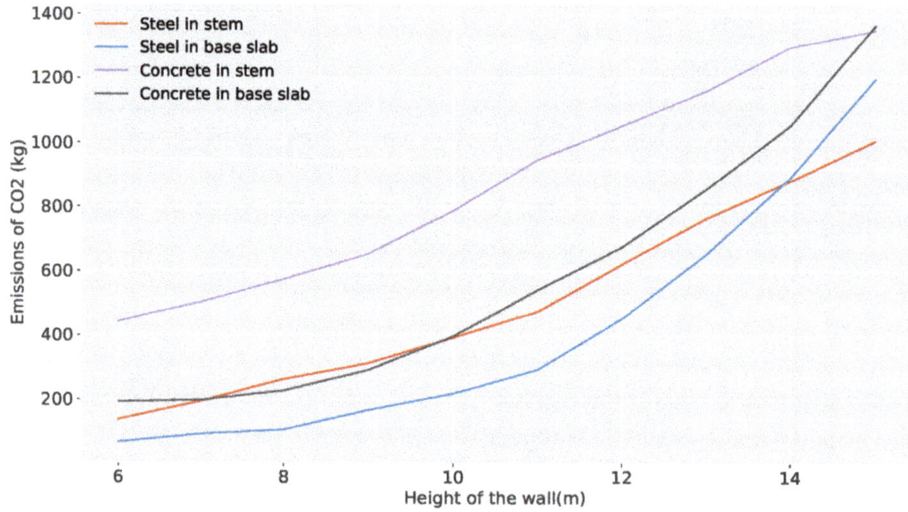

Figure 8. Emissions of concrete and steel in the stem and foundation for CO_2 optimization.

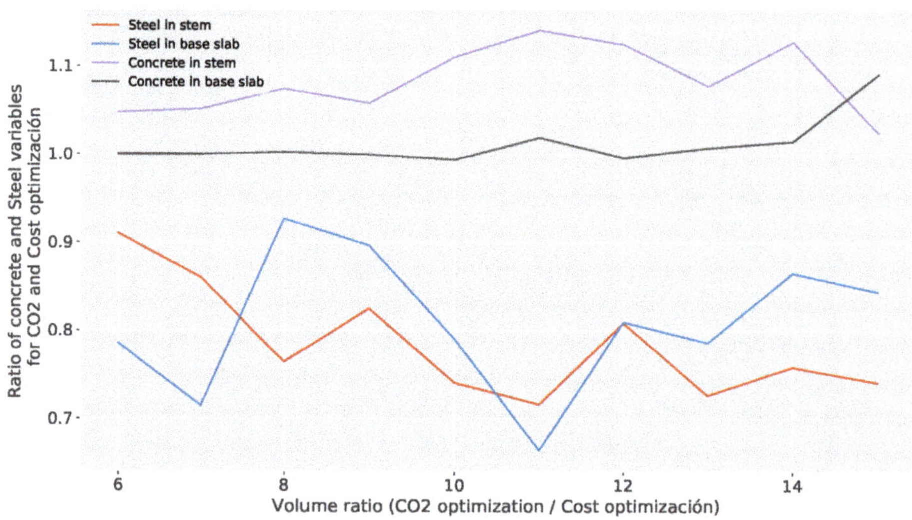

Figure 9. Ratio of concrete and steel variables.

Table 5. Optimal costs and emissions for concrete and steel variables.

	Cost Optimization				Emissions Optimization			
Height	Steel in Stem (Kg)	Steel in Base (Kg)	Concrete in Stem (m^3)	Concrete in Base (m^3)	Steel in Stem (Kg)	Steel in Base (Kg)	Concrete in Stem (m^3)	Concrete in Base (m^3)
6	53.27	30.51	1.88	0.86	48.48	23.96	1.97	0.86
7	79.91	45.90	2.13	0.88	68.73	32.80	2.24	0.88
8	121.48	40.39	2.38	1.01	92.81	37.42	2.55	1.01
9	134.06	66.03	2.75	1.30	110.55	59.20	2.91	1.30
10	187.74	98.09	3.16	1.78	139.04	77.53	3.51	1.77
11	234.57	156.61	3.70	2.37	167.98	103.98	4.21	2.41
12	272.33	197.82	4.17	3.02	220.12	160.13	4.69	3.01
13	372.58	293.64	4.80	3.78	270.65	230.85	5.16	3.80
14	411.92	365.27	5.13	4.63	312.40	316.09	5.77	4.69
15	481.51	503.15	5.87	5.58	356.85	424.82	6.02	6.09

4.3. Algorithm Comparison

This section aims to compare the results obtained by the BH algorithm with the results published in [10] and [13]. In these works, the harmony search algorithm was used to optimize a counterfort retaining wall. To make the comparison, we analyzed the best solution obtained for each of the different heights of the wall, in addition to comparing the distribution of the total solutions obtained for the different heights.

Harmony search (HS) is a swarm intelligence algorithm that was proposed in [27]. One of the important characteristics for which it was selected is that it allows the selection of features in discrete search spaces. For the study of earth-retaining wall buttresses, a variant of HS was used that additionally incorporates threshold acceptance, and was proposed in [45].

In Figure 10a, the best results for the BH and HS algorithms are compared. In the comparison, all parameters were kept constant except for height. From Figure 10a, it is observed that for small heights of the wall, the cost of the solutions obtained by both algorithms is similar. As the wall grows in height, the curves have a greater separation at the height of 14 m, which obtains the greatest difference, of 4.74% in favor of BH. In the case of optimization of CO_2 emissions, in Figure 10b, the curves have behavior similar to that in the case of cost optimization. For small values of the wall height, very similar values are obtained and, in particular, at the height of 6 m, HS is greater than BH by 0.72%. As the height of the wall grows, the quality of BH solutions improves with respect to HS again at 14 m, where the greatest difference of 4.67% is obtained.

To compare the solutions, we use violin plots. To make the comparison, we will normalize the points using the best values obtained independently of the algorithm and using the Equation (12).

$$Gap = 100 * \frac{value - bestvalue}{bestvalue} \quad (12)$$

The comparison between the BH and HS solutions in the case of costs is shown in Figure 11a. In this Figure, it is observed, in the case of BH, it obtains better interquartile ranges than in HS for all heights. The dispersion of the solutions is very similar in both algorithms. From the height of 12 m onwards, the dispersion increases significantly in both algorithms. The solutions obtained from the optimization of CO_2 emissions are shown in Figure 11b. Again, BH has a better interquartile range than HS, except for the height of 6 m. The increase in dispersion in both algorithms is observed from height of 12 m onwards.

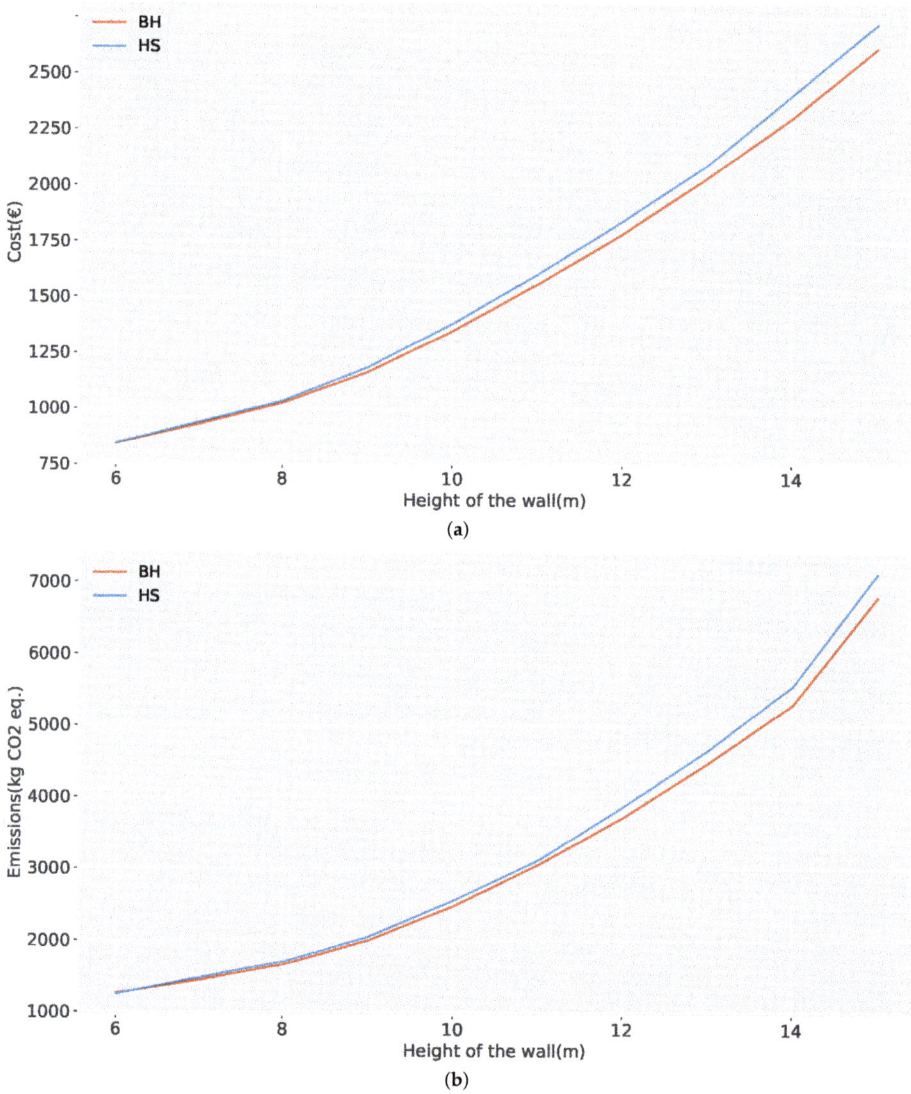

Figure 10. Best solution comparison between BH and HS. (**a**) Cost (€); (**b**) Emissions (kg CO2 eq.).

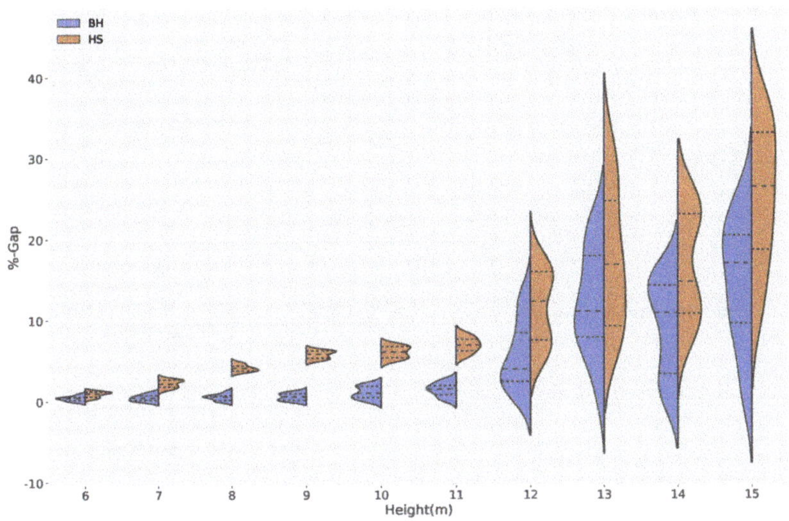

(a)

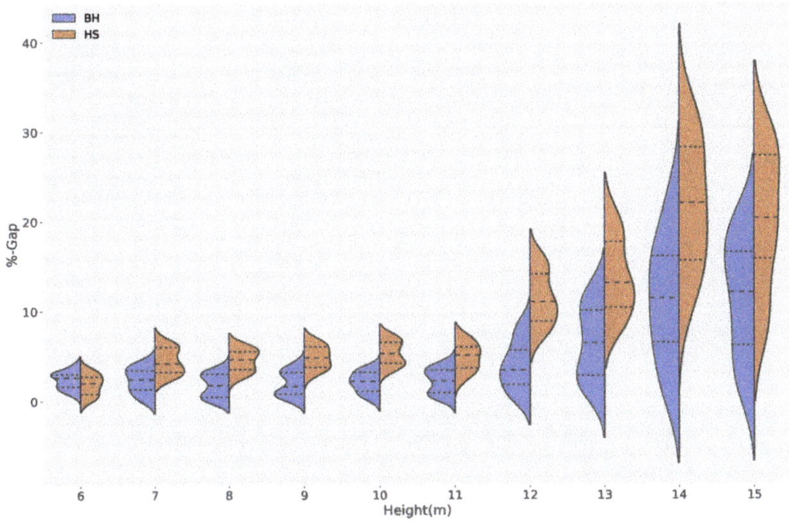

(b)

Figure 11. Comparison between BH and HS results. (**a**) % Cost gap; (**b**) % Emissions gap.

5. Conclusions

This article studies a parametric optimization of a buttressed earth-retaining wall using a discrete black hole algorithm. The analysis was developed considering two objective functions: One function that optimizes the cost of the structure and another that minimizes CO_2 emissions. In the different experiments, the height of the wall was varied, and we subsequently compared the results of both optimizations. Working with the target of reducing emissions provides stabilized results while maintaining the economic target. The results obtained with the BH heuristic also suggest that there is a relationship between cost and equivalent CO_2 emissions. For small heights of 6 and 7 m, the optimal CO_2 emission and structure costs do not deviate much from each other. As the wall grows, this difference increases, reaching a maximum at 11 m, where the emission deviates 6.16% and the cost 7.49%. When we analyze the main cause of this difference, it is found that it is related to the use of steel and cement. Solutions that minimize emissions prefer to use more concrete and less steel than those that optimize cost. Regarding the design variables, for optimal solutions, most of the values are very similar in both optimizations, with the exception of the variable of distance between buttresses, whose emission optimization value is always less than that obtained from cost optimization. For high heights (15 m), there is an inversion between the toe and heel length variables. As a result of the second analysis, in both optimizations and for the different wall heights, the concrete and the resulting steel in all cases are the C25/30 and BS500. When analyzing the dispersion results of the solutions for the different heights of the wall, it is observed that the dispersion of the solutions (Figure 11a,b) increases. This increase is considerable from the height of 12 m onwards. The above is an indication that our optimization problem is becoming more difficult as the height of the wall increases. This is consistent with the fact that at higher heights, it is more difficult to obtain stability of the wall concerning overturning and sliding. Finally, when we compare BH with HS, we observe that as we increase the height, BH performs better than HS, arriving at the height of 14 m at a difference of 4.67% in favor of BH in the optimization of emissions and 4.74% in cost minimization.

As a new line of research, it is interesting to explore robust optimization. Consider the optimization of the wall, but incorporating different values of some of the design variables within the optimization, as well as edge conditions, such as natural disasters. This makes the optimization problem much more difficult to address. Aiming at the latter, hybrid techniques can be used where metaheuristic algorithms are integrated with machine learning [46,47] techniques in order to improve the quality and convergence time of the solutions. This allows the tackling of more difficult problems, such as robust optimization in reasonable execution times.

Author Contributions: Conceptualization, V.Y., J.V.M. and J.G.; methodology, V.Y., J.V.M. and J.G.; software, J.V.M. and J.G.; validation, V.Y., J.V.M. and J.G.; formal analysis, V.Y.; investigation, J.V.M. and J.G.; data curation, J.V.M.; writing–original draft preparation, J.G.; writing–review and editing, V.Y., J.V.M. and J.G.;funding acquisition, V.Y. and J.G. All authors have read and agreed to the published version of the manuscript.

Funding: The authors acknowledge the financial support of the financial support of the Spanish Ministry of Economy and Competitiveness, along with FEDER funding (Project: BIA2017-85098-R) to the first and second authors, and the Grant CONICYT/FONDECYT/INICIACION/11180056 to the third author.

Conflicts of Interest: The authors declare no conflict of interest.

References

1. Frangopol, D.M. Life-cycle performance, management, and optimisation of structural systems under uncertainty: Accomplishments and challenges. *Struct. Infrast. Eng.* **2011**, *7*, 389–413. [CrossRef]
2. Ramesh, T.; Prakash, R.; Shukla, K.K. Life cycle energy analysis of buildings: An overview. *Energy Build.* **2010**, *42*, 1592–1600. [CrossRef]
3. Boesch, M.E.; Hellweg, S. Identifying improvement potentials in cement production with life cycle assessment. *Environ. Sci. Technol.* **2010**, *44*, 9143–9149. [CrossRef]
4. Serpell, A.; Kort, J.; Vera, S. Awareness, actions, drivers and barriers of sustainable construction in Chile. *Technol. Econ. Dev. Econ.* **2013**, *19*, 272–288. [CrossRef]
5. Yusof, N.A.; Abidin, N.Z.; Zailani, S.H.M.; Govindan, K.; Iranmanesh, M. Linking the environmental practice of construction firms and the environmental behaviour of practitioners in construction projects. *J. Clean. Prod.* **2016**, *121*, 64–71. [CrossRef]
6. Wang, T.; Lee, I.S.; Kendall, A.; Harvey, J.; Lee, E.B.; Kim, C. Life cycle energy consumption and GHG emission from pavement rehabilitation with different rolling resistance. *J. Clean. Prod.* **2012**, *33*, 86–96. [CrossRef]
7. Wang, E.; Shen, Z. A hybrid Data Quality Indicator and statistical method for improving uncertainty analysis in LCA of complex system—Application to the whole-building embodied energy analysis. *J. Clean. Prod.* **2013**, *43*, 166–173. [CrossRef]

8. Barandica, J.M.; Fernandez-Sanchez, G.; Berzosa, A.; Delgado, J.A.; Acosta, F.J. Applying life cycle thinking to reduce greenhouse gas emissions from road projects. *J. Clean. Prod.* **2013**, *57*, 79–91. [CrossRef]
9. Aguado, A.; Caño, A.D.; de la Cruz, M.P.; Gomez, D.; Josa, A. Sustainability assessment of concrete structures within the Spanish structural concrete code. *J. Constr. Eng. Manag.* **2012**, *138*, 268–276. [CrossRef]
10. Molina-Moreno, F.; García-Segura, T.; Martí, J.V.; Yepes, V. Optimization of counterfort retaining walls using hybrid harmony search algorithms. *Eng. Struct.* **2017**, *134*, 205–216. [CrossRef]
11. Yepes, V.; Martí, J.V.; García-Segura, T. Cost and CO_2 emission optimization of precast–prestressed concrete U-beam road bridges by a hybrid glowworm swarm algorithm. *Autom. Constr.* **2015**, *49*, 123–134. [CrossRef]
12. Worrell, E.; Price, L.; Martin, N.; Hendriks, C.; Meida, L.O. Carbon dioxide emissions from the global cement industry. *Ann. Rev. Energy Environ.* **2001**, *26*, 303–329. [CrossRef]
13. Molina-Moreno, F.; Martí, J.V.; Yepes, V. Carbon embodied optimization for buttressed earth-retaining walls: Implications for low-carbon conceptual designs. *J. Clean. Prod.* **2017**, *164*, 872–884. [CrossRef]
14. Yepes, V.; Gonzalez-Vidosa, F.; Alcala, J.; Villalba, P. CO_2-optimization design of reinforced concrete retaining walls based on a VNS-threshold acceptance strategy. *J. Comput. Civ. Eng.* **2012**, *26*, 378–386. [CrossRef]
15. Yoon, Y.C.; Kim, K.H.; Lee, S.H.; Yeo, D. Sustainable design for reinforced concrete columns through embodied energy and CO_2 emission optimization. *Energy Build.* **2018**, *174*, 44–53. [CrossRef]
16. Sierra, L.A.; Pellicer, E.; Yepes, V. Social sustainability in the life cycle of Chilean public infrastructure. *J. Constr. Eng. Manag.* **2016**, *142*, 05015020. [CrossRef]
17. Sierra, L.A.; Yepes, V.; García-Segura, T.; Pellicer, E. Bayesian network method for decision-making about the social sustainability of infrastructure projects. *J. Clean. Prod.* **2018**, *176*, 521–534. [CrossRef]
18. Moayyeri, N.; Gharehbaghi, S.; Plevris, V. Cost-Based Optimum Design of Reinforced Concrete Retaining Walls Considering Different Methods of Bearing Capacity Computation. *Mathematics* **2019**, *7*, 1232. [CrossRef]
19. Pons, J.J.; Penadés-Plà, V.; Yepes, V.; Martí, J.V. Life cycle assessment of earth-retaining walls: An environmental comparison. *J. Clean. Prod.* **2018**, *192*, 411–420. [CrossRef]
20. Zastrow, P.; Molina-Moreno, F.; García-Segura, T.; Martí, J.V.; Yepes, V. Life cycle assessment of cost-optimized buttress earth-retaining walls: A parametric study. *J. Clean. Prod.* **2017**, *140*, 1037–1048. [CrossRef]
21. Lee, D.; Kang, G.; Nam, C.; Cho, H.; Kang, K.I. Stochastic Analysis of Embodied Carbon Dioxide Emissions Considering Variability of Construction Sites. *Sustainability* **2019**, *11*, 4215. [CrossRef]
22. Amirkhani, S.; Bahadori-Jahromi, A.; Mylona, A.; Godfrey, P.; Cook, D. Impact of low-e window films on energy consumption and CO_2 emissions of an existing UK hotel building. *Sustainability* **2019**, *11*, 4265. [CrossRef]
23. Lu, K.; Jiang, X.; Tam, V.W.; Li, M.; Wang, H.; Xia, B.; Chen, Q. Development of a Carbon Emissions Analysis Framework Using Building Information Modeling and Life Cycle Assessment for the Construction of Hospital Projects. *Sustainability* **2019**, *11*, 6274. [CrossRef]
24. De Medeiros, G.F.; Kripka, M. Optimization of reinforced concrete columns according to different environmental impact assessment parameters. *Eng. Struct.* **2014**, *59*, 185–194. [CrossRef]
25. Hussain, K.; Salleh, M.N.M.; Cheng, S.; Shi, Y. Metaheuristic research: A comprehensive survey. *Artif. Intell. Rev.* **2019**, *52*, 2191–2233. [CrossRef]
26. Bozorg-Haddad, O. (Ed.) *Advanced Optimization by Nature-Inspired Algorithms;* Springer: Singapore, 2018.
27. Geem, Z.W.; Kim, J.H.; Loganathan, G.V. A new heuristic optimization algorithm: Harmony search. *Simulation* **2001**, *76*, 60–68. [CrossRef]
28. Kirkpatrick, S.; Gelatt, C.D.; Vecchi, M.P. Optimization by simulated annealing. *Science* **1983**, *220*, 671–680. [CrossRef]
29. Cerny, V. Thermodynamical approach to the traveling salesman problem: An efficient simulation algorithm. *J. Optim. Theory Appl.* **1985**, *45*, 41–51. [CrossRef]
30. Carbonell, A.; González-Vidosa, F.; Yepes, V. Design of reinforced concrete road vaults by heuristic optimization. *Adv. Eng. Softw.* **2011**, *42*, 151–159. [CrossRef]
31. Zavadskas, E.K.; Antucheviciene, J.; Vilutiene, T.; Adeli, H. Sustainable decision-making in civil engineering, construction and building technology. *Sustainability* **2018**, *10*, 14. [CrossRef]
32. Shi, X.; Wu, L.; Meng, X. A new optimization model for the sustainable development: Quadratic knapsack problem with conflict graphs. *Sustainability* **2017**, *9*, 236. [CrossRef]
33. Sierra, L.A.; Yepes, V.; Pellicer, E. A review of multi-criteria assessment of the social sustainability of infrastructures. *J. Clean. Prod.* **2018**, *187*, 496–513. [CrossRef]

34. García, J.; Moraga, P.; Valenzuela, M.; Crawford, B.; Soto, R.; Pinto, H.; Pena, A.; Altimiras, F.; Astroga, G. A Db-Scan Binarization Algorithm Applied to Matrix Covering Problems. *Comput. Intell. Neurosci.* **2019**, *2019*, 16. [CrossRef] [PubMed]
35. García, J.; Crawford, B.; Soto, R.; Astorga, G. A clustering algorithm applied to the binarization of swarm intelligence continuous metaheuristics. *Swarm Evol. Comput.* **2019**, *44*, 646–664. [CrossRef]
36. García, J.; Crawford, B.; Soto, R.; Astorga, G. A percentile transition ranking algorithm applied to binarization of continuous swarm intelligence metaheuristics. In Proceedings of the International Conference on Soft Computing and Data Mining, Johor, Malaysia, 6–8 February 2018; pp. 3–13.
37. Yepes, V.; Alcala, J.; Perea, C.; González-Vidosa, F. A parametric study of optimum earth-retaining walls by simulated annealing. *Eng. Struct.* **2008**, *30*, 821–830. [CrossRef]
38. García-Segura, T.; Yepes, V.; Alcalá, J. Life-cycle greenhouse gas emissions of blended cement concrete including carbonation and durability. *Int. J. Life Cycle Assessment* **2014**, *19*, 3–12. [CrossRef]
39. Ministerio de Fomento. *EHE: Code of Structural Concrete*; Ministerio de Fomento: Madrid, Spain, 2008.
40. Ministerio de Fomento. *CTE. DB-SE. Structural Safety: Foundations*; Ministerio de Fomento: Madrid, Spain, 2007. (In Spanish)
41. Huntington, W.C. *Earth Pressures and Retaining Wall*; John Wiley and Sons: New York, NY, USA, 1957.
42. Calavera, J. *Muros de Contención y Muros de Sótano*; Intemac: Madrid, Spain, 2001. (In Spanish)
43. CEB-FIB. Model Code. In *Design Code*; Thomas Telford Services Ltd.: London, UK, 2001.
44. Hatamlou, A. Black hole: A new heuristic optimization approach for data clustering. *Inf. Sci.* **2013**, *222*, 175–184. [CrossRef]
45. García-Segura, T.; Yepes, V.; Alcalá, J.; Pérez-López, E. Hybrid harmony search for sustainable design of post-tensioned concrete box-girder pedestrian bridges. *Eng. Struct.* **2015**, *92*, 112–122. [CrossRef]
46. García, J.; Lalla-Ruiz, E.; Voß, S.; López Droguett, E. Enhancing a machine learning binarization framework by perturbation operators: Analysis on the multidimensional knapsack problem. *Int. J. Mach. Learn. Cyber.* **2020**, doi:10.1007/s13042-020-01085-8. [CrossRef]
47. García, J.; Crawford, B.; Soto, R.; Castro, C.; Paredes, F. A k-means binarization framework applied to multidimensional knapsack problem. *Appl. Intell.* **2018**, *48*, 357–380. [CrossRef]

sustainability

Article

Comparative Analysis of Economic Impacts of Sustainable Vertical Extension Methods for Existing Underground Spaces

Soo-yeon Seo [1], Byunghee Lee [2] and Jongsung Won [1,*]

[1] School of Architecture, Korea National University of Transportation, Chungju 27389, Korea; syseo@ut.ac.kr
[2] Korea Construction Technology Consulting & Eng. Co., Ltd., Seoul 05548, Korea; plus500@daum.net
* Correspondence: jwon@ut.ac.kr

Received: 14 January 2020; Accepted: 25 January 2020; Published: 29 January 2020

Abstract: Without demolishing an entire existing building, it is possible to sustainably expand its underground spaces to enhance the building's functionality. However, there have been a few relevant studies exploring this option, and they did not consider the financial feasibilities of underground vertical extension methods. Therefore, this paper analyzes the economic impacts of three sustainable vertical extension methods for existing underground spaces. The extension methods were the (1) bottom-up, (2) normal top-down, and (3) top-down with multi-post downward (MPD) methods. In order to analyze and compare the economic impacts of the underground vertical extension methods, 24 illustrative examples were generated in this paper. Construction costs of the three sustainable vertical extension methods for existing underground spaces are calculated and compared. Those are based on the quantity of used materials in the construction phase and dismantled materials in the demolition phase, as well as unit costs of each material. In addition, the structural stabilities of the examples are analyzed using MIDAS Gen 2017. As the results, the top-down method with MPD was the lowest sustainable method for vertically expanding underground spaces compared to other two methods under the same condition. Moreover, the higher the number of underground floors of existing buildings and the greater the number of extended basement floors, the more economically advantageous was the top-down method with MPD. Considering their structural stabilities and economic impacts of the extension methods help practitioners to select appropriate construction techniques and reduce costs, risks, and the amount of generated construction and demolition waste.

Keywords: vertical extension method; underground; existing building; economic impact

1. Introduction

The use of underground space in buildings in congested urban areas has been increasing since there is a lack of available aboveground space [1–4] and the development of additional underground space in existing cities can provide new potential for urban development [5]. Underground structures have significant impacts on the environment [4,6] and underground space can reduce environmental contamination and improve the quality of the environment [2,7]. Regardless of stability of existing buildings, sometimes buildings have been demolished and newly constructed due to the lack of underground parking spaces or aging of mechanical, electrical, and plumbing (MEP) facilities [8]. Excessive demolition and new construction of existing buildings increase the amount of generated construction and demolition (C&D) waste [8]. Without demolishing an entire building, it is possible to sustainably expand its underground spaces vertically and horizontally to enhance the functionality. It can be an opportunity to reduce the amount of generated C&D waste and create sustainable urban development in existing cities. However, there are relatively few cases of vertically expanding underground spaces in buildings because of several challenges, such as the difficulty of risk management related to construction of underground structures and minimization of structural effects on ground floors of existing and surrounding buildings. To address the challenges of vertical extensions for underground spaces in existing buildings, a few previous approaches have been conducted: (1) practical case studies on vertical extensions of underground spaces in existing buildings [9–12], (2) development of a connection system between existing piles and new piles installed for extended underground spaces [13], and (3) development of vertical underground extension processes for existing buildings [14–16]. Bing [10] introduced an application case of a floating underground extension method for expanding parking lot spaces in residential buildings, where the floating extension method excavates the area underneath the existing building to extend it vertically downward without destroying the building. They also derived factors influencing the extension planning and structural stability of existing buildings. Kim, Bang, and Lim [13] proposed a system to connect piles installed in existing underground spaces with foundations newly constructed for extended underground spaces. It helped secure structural stability and reduce the number of reinforcement piles. However, the construction processes were to be complicated, because temporary micro-piles and frames should be installed during excavation and dismantled for the vertical underground extension in the proposed connection system. Park, Lew, Choi, and Lee [14] introduced a floating underground space extension method applied to actual sites for the preservation and expansion of cultural-heritage buildings. After excavation, double-tube micro-piles and jack-up systems were used for supporting the buildings. Kim, Lee, Kim, Koo, Jung, and Seo [15] and Jung, Kim, Lee, Hwang, and Seo [16] proposed a new conceptual construction process involving floating methods for expanding existing buildings and conceptually arranged the process for selecting the most suitable method among various construction methods according to characteristics of each site. Selection processes for appropriate construction techniques can influence both the time and costs of construction, as well as the environment [17].

Sustainable vertical extension methods for underground spaces without demolishing existing buildings may vary depending on site conditions of architecture, engineering, and construction (AEC) projects. However, previous studies did not consider detailed excavation and construction processes based on various site conditions. In particular, they did not analyze the economic feasibility of the proposed underground vertical extension methods for existing buildings. Therefore, a process to analyze economic impacts of three sustainable vertical extension methods of existing underground spaces is proposed based on the quantity of used or demolished materials on site. The three sustainable methods are bottom-up, normal top-down, and top-down multiple posts downward (MPDs) methods. To analyze their economic impacts, construction costs of 24 illustrative examples to which three sustainable vertical extension methods for existing underground spaces are applied, are calculated, and compared. In addition, their structural stabilities are also analyzed using MIDAS Gen 2017.

The structure of this paper is as follows. Section 2 describes the materials and research methods used in this paper. Section 3 includes processes and details of three sustainable vertical extension methods for underground spaces in existing buildings. Section 4 describes the overview of the 24 illustrative examples. Section 5 includes results of the structural analyses with the three sustainable vertical extension methods for underground spaces in existing buildings. In Section 6, the economic impacts of the three sustainable vertical extension methods for underground spaces in the illustrative examples are analyzed, followed by discussion and conclusions in Section 7.

2. Materials and Methods

Figure 1 shows an overall research process of this paper which analyzes structural stabilities and economic impacts of the three sustainable vertical extension methods of existing underground spaces. The three sustainable methods considered in this paper are as follows:

(1) Bottom-up method: the construction of floor structures to be extended are carried out after finishing all of the excavation of the soil under the existing building.

(2) Normal top-down method: the construction of each floor structure to be extended is carried out after finishing the excavation of soil to just below one floor under the existing building and the process is repeated up to the construction of the bottom.
(3) Top-down with multiple posts downward (MPDs) method: the construction procedure is the same as that of the normal top-down method but it uses multiple posts for supporting the existing building.

Vertically expanding underground spaces of existing building with basement floors is a complex task since many perspectives, such as economic, structural stability, and environmental perspectives should be considered. The details and processes of the three sustainable extension methods are described in Section 3. In order to analyze the economic impacts of the three sustainable methods, 24 illustrative examples were generated through considering the number of basement floors of existing buildings, the number of basement floors that are vertically extended, and whether demolition works are included or not. Prior to economic analyses of the three underground vertical extension methods, their structural stabilities should be reviewed. The structural stabilities of the illustrative examples were analyzed using MIDAS Gen 2017 in this paper. The reliability of the program was verified earlier by comparing it with other analysis programs [18]. Currently, the program is used for structural analysis of almost all buildings in South Korea.

The calculated vertical underground extension cost of the 24 illustrative examples consisted of material and labor costs. Material costs were classified as reinforcement, demolition, earthworks, structural frame construction, and finishing costs. The material cost of each work type was calculated by multiplying the quantity and unit price of materials required for the vertical extension of an underground space. The quantity of materials was estimated by an expert considering the size, the number, and position of all members assumed for the structural analyses. The working experience of the experts who participated in the process was approximately 20 years. The labor cost is generally assumed as 15% of total material cost for underground construction works in South Korea. Based on calculated material and labor costs of each case, their economic impacts were analyzed and compared.

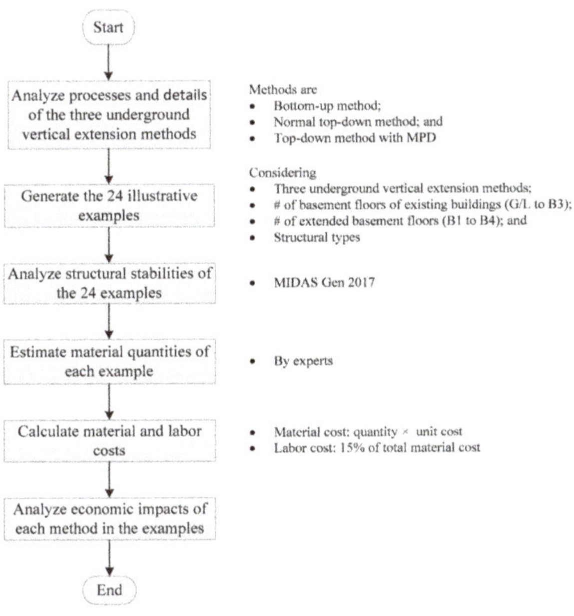

Figure 1. The research process.

3. Sustainable Vertical Extension Methods for Underground Spaces in Existing Buildings

In this paper, the three sustainable vertical extension methods used for underground spaces in existing buildings were the bottom-up, normal top-down, and top-down with MPDs methods. In selecting a vertical extension method for this paper, two conditions were considered: (1) a building has more than one basement floor, and (2) an existing underground structure can be demolished or reused for vertical extensions. The normal top-down and bottom-up methods can be used for vertically expanding underground spaces after demolishing existing underground structures [19]. The top-down method with MPD can expand underground space without demolishing existing underground structures.

The top-down method, which is widely used in congested AEC projects, helps secure the safety of retaining walls and work spaces and reduce construction duration [1,20]. To apply the top-down method for sustainable vertical extension of existing underground spaces, columns should be installed in the ground prior to excavation. In most cases, the basement floors of existing buildings are generally demolished prior to the installation of columns to enhance constructability. A percussion rotary drill (PRD) is commonly used for installing columns. However, a PRD has large diameter and is relatively expensive. Therefore, it is not recommended for vertical extension of underground spaces if an existing building has less than three basement floors; normally the existing floors of a South Korean residential building include one or two floors for parking lots. On the other hand, the MPD method uses augur equipment, so that it is possible to use a small diameter post such as micro piles. If the installed MPDs have sufficient resistance to support the existing building without any additional supports during the excavation, the top-down method with MPDs can be used without the removal of existing underground structures. In particular, the top-down method with MPDs is more efficient for vertically downward expanding underground spaces in urban areas because of the lack of available spaces for construction and demolition. The installed temporary multiple posts could be reused after disassembly. The bottom-up method is a common construction technique. It includes an open-cut method with struts to install retaining walls. It is easy to trigger the collapse of retaining walls or cracks in the structures of surrounding buildings because of settlement of the surrounding ground. Therefore, a careful design is required to avoid stress concentration on a part during the dismantlement of temporary struts. The top-down method of constructing the structure of building along with the excavation can be a relatively safe construction.

To select an appropriate vertical extension method for the site with existing underground structure, relevant drawings and geological conditions on site should be carefully reviewed [8]. Figure 2 shows a process to select an appropriate vertical extension method based on the several site conditions. The net distance between the basement outer wall of the targeted existing building and the outer wall of the adjacent building should be at least 1.2 m [21]. Otherwise, the distance is too short to construct an earth retaining wall between these buildings. In this case, the earth retaining wall should be installed inside the structure of the existing building. Consequently, the useful area of underground spaces after vertical underground extension is to be reduced. In this paper, the net distance between two walls is assumed to be more than 1.2 m. The processes and details of the three vertical extension methods for existing underground spaces are described in the following sub-sections.

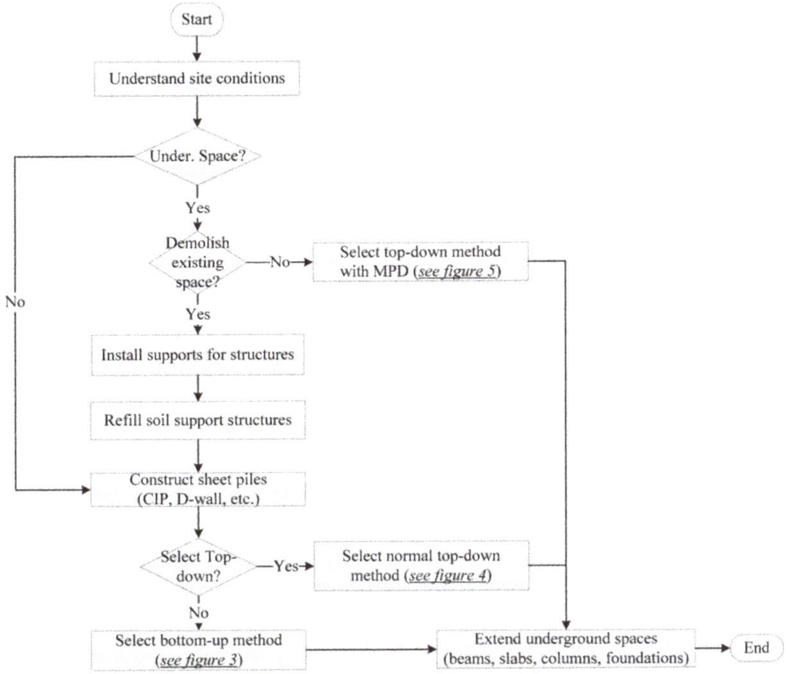

Figure 2. The process to select an appropriate vertical extension method for underground spaces in existing buildings by considering site conditions.

3.1. Bottom-Up Method with Struts

The vertical underground extension using the bottom-up method constructs structural frames after completing excavation to foundation levels using temporary struts. Figure 3 represents the construction process of the bottom-up method with struts, from the first basement level to the third basement level. Underground spaces were extended vertically after demolishing the basements of the existing building. Before demolishing, the existing structure needed to be reinforced with temporary members to support the weight of the equipment for installing the new retaining walls (Figure 3b). After Installing retaining walls, temporary struts and H-piles were installed to provide horizontal and vertical structural stability for the underground space during the dismantling of the existing structure and the excavation for an extension. The vertical interval of the struts was 2.5 m. Structural frames for the underground space were constructed from the lowest bottom level to the ground level. After the structural frame of a floor was completed, the temporary members installed on the floor such as struts were removed, as shown in Figure 3e.

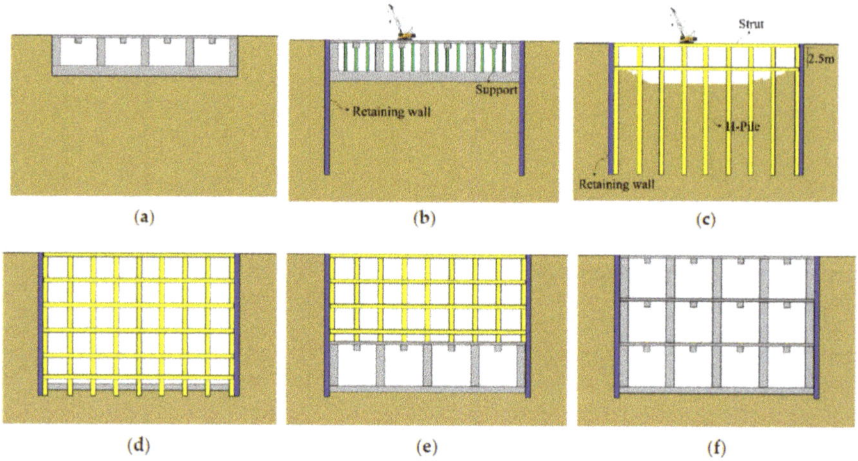

Figure 3. Vertical extension process for existing underground spaces using the bottom-up method (from a building with one basement floor to three basement floors). The process is listed as: (**a**) Existing underground structures before works; (**b**) Installing of earth retaining walls after reinforcing the existing structure; (**c**) Construction of temporary struts and H-piles after demolition of the existing structure; (**d**) Foundation construction after Installing of struts up to base; (**e**) Construction of a structural frame (from B3 to B1) and dismantling struts; and (**f**) Completion.

3.2. Normal Top-Down Method

The vertical extension using the normal top-down method constructs structural frames sequentially from the first basement level to the lowest bottom level [19]. The constructed frames, such as slabs and beams, support earth retaining walls instead of temporary struts [15]. Furthermore, construction costs can be reduced because laborers can work on the constructed slabs [15]. However, it is difficult to excavate and install formworks (within the narrow space) under the constructed slab [15]. As bottom-up method with struts, the existing structure needed to be reinforced with temporary members for supporting the weight of the equipment for installing the new retaining walls before demolishing. Struts should be partially arranged to ensure structural safety against earth pressure during the dismantling of existing structures. Figure 4 shows the process of extending the existing underground spaces vertically, from the first basement level to the third basement level, using the normal top-down method while demolishing the existing basement floors.

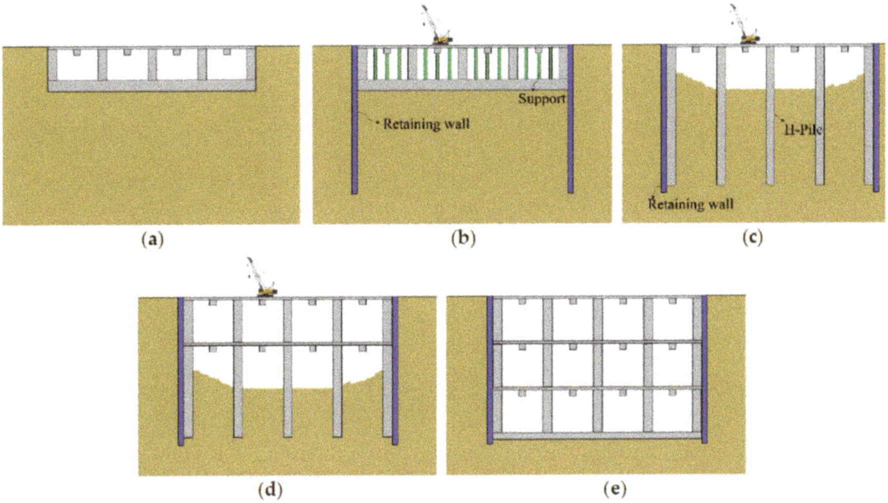

Figure 4. Vertical extension process for existing underground spaces using the normal top-down method (from a building with one basement floor to three basement floors). The process is listed as: (**a**) Existing underground spaces before works; (**b**) Installing of earth retaining walls after reinforcing existing structure; (**c**) Demolition of basement Level 1 and structural frame construction (columns) after installing earth retaining walls; (**d**) Excavation and construction of structural frame (slabs and beams) (from B2 to B3); and (**e**) Completion.

3.3. Top-Down Method with MPDs [22]

Figure 5 represents the process of vertically expanding existing underground spaces without demolishing existing structures, from the first basement level to the third basement level, using the top-down method with MPDs. To extend underground space vertically after reinforcing existing underground structure without its demolition, several pieces of equipment were placed at the top of the existing underground structure in order to install additional earth retaining walls. Therefore, prior to installation of the earth retaining walls, the supports for the structural stability of the existing slabs were installed and reinforced by considering the loads from the equipment for the earthwork (Figure 5b). To prevent the collapse of columns in the basement of the existing building during the excavation, it was necessary to install multiple posts, such as micro piles, around the columns, as shown in Figure 5c. The four micro piles installed on the outside of one column were slender, resulting in the buckling behavior. Therefore, it was necessary to tie up those to one set and connect several sets of micro piles to reduce the risk of buckling failure. Then, the foundation of the existing building was demolished, and excavation work was carried out from the bottom level of the existing building up to the extended bottom level. After completing the installation of the new foundations and columns located in the extended underground structure, as shown in Figure 5d,e, the temporary micro piles were removed.

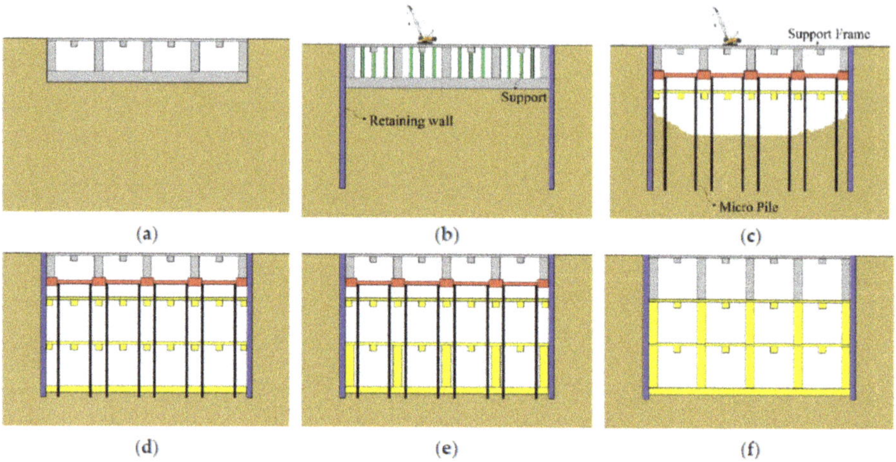

Figure 5. Vertical extension process for existing underground spaces using the top-down method with the MPD method (from a building with one basement floor to three basement floors). The process is listed as: (**a**) Existing underground structure; (**b**) Installation of temporary supports and retaining walls; (**c**) Installation of temporary piles and support frames and excavation (from B2 to B3); (**d**) Structural frame (beams and slabs) and foundation construction; (**e**) Installation of columns on floors (from B3 to B2); and (**f**) Decomposition of temporary piles and support frames.

4. Overview of Illustrative Examples

A rectangular-shaped underground structure was used to generate illustrative examples in this paper. The effects of the shape were not considered to simplify comparative analysis of the economic impacts corresponding to sustainable vertical extension methods of existing underground spaces. The structure used for the illustrative examples has a floor plan of 74.0 × 29.6 m (Figures 6 and 7), with a floor height of 3.4 m at each basement.

To analyze the differences among the extension costs of the three sustainable vertical extension methods for underground space according to the number of basement floors of the existing building, the illustrative examples were categorized as four types: the existing building (1) without basement floors, (2) with one basement floor, (3) with two basement floors, and (4) with three basement floors. As shown Table 1, 24 illustrative examples considering sustainable vertical extension methods, the number of basement floors of the existing building (from G/L to B3), the number of extended basement floors (from B1 to B4), and structural types were generated. If there were no basement floors of the existing building before the extension, the bottom-up method and normal top-down method were applied without demolition work. If there were existing basement floors, the bottom-up method after demolishing existing underground spaces, the normal top-down method with Percussion Rotary Drill (PRD) after demolishing existing underground spaces, and the top-down with MPDs method without demolishing existing underground spaces were compared. The normal top-down method with PRD and the reverse circulation drill (RCD) and bottom-up method were not allowed to vertically extend the underground space of an existing building with more than three basement floors. Therefore, the economic impact of only the top-down with MPDs method for sustainable vertical underground extension of an existing building with three basement floors was analyzed. The structure types that the bottom-up method, normal top-down method, and top-down method with MPD analyzed were reinforced concrete (RC), steel frame, and steel frame, respectively.

Table 1. Basic characteristics of the illustrative examples.

No.	Number of Basement Floors before Extension	Whether to Dismantle the Existing Structure	Number of Basement Floors after Extension	Total Height of Basement Floors (m)	Vertical Extension Method for Underground Spaces	Structural Type
Case 1	0		1	3.4	Bottom-up	RC
Case 2	0		1	3.4	Top-down	Steel
Case 3	0		2	6.8	Bottom-up	RC
Case 4	0		2	6.8	Top-down	Steel
Case 5	0		3	10.2	Bottom-up	RC
Case 6	0		3	10.2	Top-down	Steel
Case 7	0		4	13.6	Bottom-up	RC
Case 8	0		4	13.6	Top-down	Steel
Case 9	1	Dismantle	2	6.8	Bottom-up	RC
Case 10	1	Dismantle	2	6.8	Top-down	Steel
Case 11	1	No dismantle	2	6.8	Top-down (MPD)	Steel
Case 12	1	Dismantle	3	10.2	Bottom-up	RC
Case 13	1	Dismantle	3	10.2	Top-down	Steel
Case 14	1	No dismantle	3	10.2	Top-down (MPD)	Steel
Case 15	1	Dismantle	4	13.6	Bottom-up	RC
Case 16	1	Dismantle	4	13.6	Top-down	Steel
Case 17	1	No dismantle	4	13.6	Top-down (MPD)	Steel
Case 18	2	Dismantle	3	10.2	Bottom-up	RC
Case 19	2	Dismantle	3	10.2	Top-down	Steel
Case 20	2	No dismantle	3	10.2	Top-down (MPD)	Steel
Case 21	2	Dismantle	4	13.6	Bottom-up	RC
Case 22	2	Dismantle	4	13.6	Top-down	Steel
Case 23	2	No dismantle	4	13.6	Top-down (MPD)	Steel
Case 24	3	No dismantle	4	13.6	Top-down (MPD)	Steel

5. Structural Analyses of the Examples

Prior to the economic analyses of the three vertical underground extension methods, structural stabilities of the illustrative examples were checked. Structural analyses were carried out in the three illustrative examples to extend underground space vertically (from the first to third basement level) using the three sustainable methods. MIDAS Gen 2017 was used for the structural analysis by phase, including demolishment of the existing B1 floor, excavation to the B4 floor, and construction of a structural frame in the underground space.

The design of methods based on structural analysis can be roughly classified into two types. The first is to ensure the safety of the whole building when the underground structure is completed. The second is to ensure structural safety for each construction phase. In the bottom-up method, since the structural frame is constructed after the completion of the excavation to the lowest floor, the safety of the entire structure was examined after completion of construction. The frame at this time was a reinforced concrete structure.

The soil condition in all cases was assumed to be sandy soil. Therefore, the effects of changes in the ground conditions were not considered in this paper. In all the illustrative examples, temporary supports were installed at 1.5-m-wide intervals in the excavated or undemolished underground space to support the existing underground space when equipment was located on the ground floor for the newly installed retaining wall. During top-down construction, the live loads acting on first and basement floor are 20.0 kN/m^2 and BF = 1.5 kN/m^2, respectively, since the first floor can be used for storage space of construction materials. The load by the materials placed on the first floor disappeared after construction. The live load acting on the first floor after construction was assumed to be 5.0 kN/m^2 which is higher than 3.0 kN/m^2 for the basement floor. The thickness of slab was 150 mm for the bottom-up method and 200 mm for the top-down methods. The load of slab finish was 2.6 kN/m^2 after construction. The earth pressure of the lowest level of strut was assumed to be 300 kN/m.

The compressive strength of concrete was 24 MPa. SD500 (yield strength: 500 MPa) and SD400 (yield strength: 400 MPa) were used for a rebar with a diameter of 16 mm or more, and a rebar less than 13 mm, respectively. SM490 (tensile strength was 490 MPa) was used for steel members. Structural design and analysis were conducted according to the Korean Building Code (KBC) 2016.

In the case of construction up to three basement levels by applying three representative methods, the size and structural analysis of each member are shown in this paper. Figures 6 and 7 show plans of the structural B1 floor and temporary struts planning in the illustrative examples, respectively. The size of beams and columns was determined by considering both the conditions required to serve as temporary members and structural members after the completion of the underground structure. Strut and H-pile sizes were determined in consideration of the requirements for use as temporary members. The ratio of forces to holding strength of girders (G1 to G4) and columns (C1 and C2) were checked. In the case of the bottom-up and top-down with MPD methods, the ratio of forces to holding strength of struts (S1 and S2) and H-piles (H1 and H2) also were checked. Temporary struts were installed at a depth of 2.5 m in both cases. Structural frames, such as constructed slabs and beams, prevent the side pressure of the earth retaining walls instead of temporary struts in the illustrative example using the normal top-down method. Therefore, structural analysis for the temporary struts was not necessary.

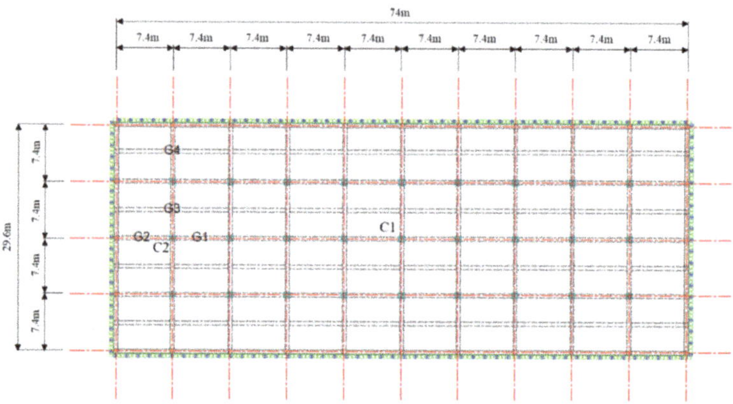

Figure 6. Structural plan of B1 floor in the illustrative examples (unit: mm).

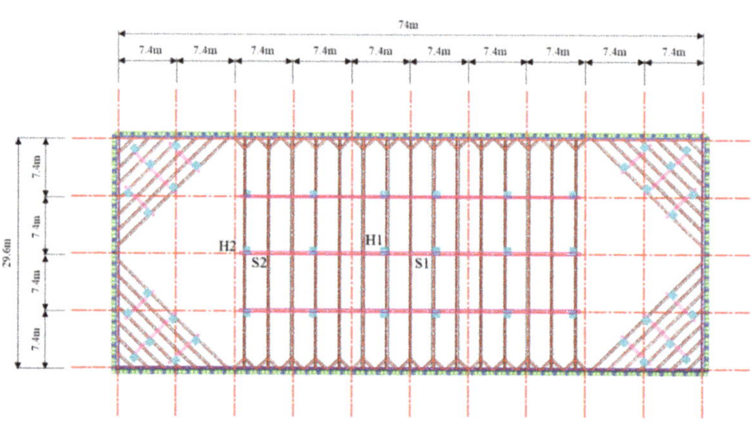

Figure 7. Temporary struts plan in the illustrative examples for the bottom-up method (unit: mm).

5.1. Bottom-Up Method

In bottom-up method, the beam and column were RC structures. The sizes of the members for Case #5 used in the vertical underground extension using the bottom-up method are described in Table 2. Sections and reinforcement details of typical girders after construction are shown in Figure 8. They were the same for each construction phase. Figure 9 shows the moment diagram of RC frame structure after the completion of construction. As the result of the structural analysis, the forces acting on the designated members of them are represented in Table 2 as a ratio to the holding strengths. In case of the bottom-up method, the conventional reinforced concrete frame system was used as structural systems after completion of excavation. The forces acting on girders of ground floor (1F) were higher than those of the basement floor (BF) because live load on 1F was higher than that on BF. Highest values were found in the main girders (G3 and G4) which were connected beams and the ratio of forces to holding strengths reached 0.95. From this, the members were shown to be optimally designed. The axial force ratios of columns C1 and C2 were 0.42 and 0.43 after construction, respectively. The strut S1 had an axial force ratio of 0.66 during construction. Consequently, it can be seen that the structural member was designed to be safe enough for the applied load.

Table 2. Member sizes and analyzed force ratios of structural members in the illustrative example using the bottom-up method (Conventional RC construction, Case #5).

Member		Size	Force Type	Ratio of Force to Holding Strength After Completion of Construction			
				1F (Center)	**1F (End)**	**BF (Center)**	**BF (End)**
Girder	G1	800 × 900	Moment	170/699 (0.24)	328/699 (0.47)	146/699 (0.21)	286/699 (0.41)
			Shear force	143/536 (0.27)	234/536 (0.44)	123/536 (0.23)	203/536 (0.38)
	G2		Moment	318/299 (0.45)	360/699 (0.52)	270/699 (0.39)	356/699 (0.51)
			Shear force	192/536 (0.36)	282/536 (0.53)	170/536 (0.32)	249/536 (0.47)
	G3		Moment	504/699 (0.72)	667/699 (0.96)	438/698 (0.63)	576/698 (0.83)
			Shear force	316/536 (0.59)	407/536 (0.76)	273/534 (0.51)	353/534 (0.66)
	G4		Moment	760/867 (0.88)	795/867 (0.92)	647/867 (0.75)	749/867 (0.86)
			Shear force	420/536 (0.78)	511/536 (0.95)	370/534 (0.69)	450/534 (0.84)
Column	C1	900 × 900	Axial force	4869/11645 (0.42)			
	C2		Axial force	5053/11645 (0.43)			
Strut	S1	H-300 × 300 × 10 × 15	Axial force	1227/1729 (0.66)			
	S2	H-300 × 150 × 6.5 × 9	Axial force	For protecting lateral buckling			
H-Pile	H1	H-300 × 300 × 10 × 15	Axial force	These members were only for the connection of temporary struts. Axial force dose was not acting on these members			
	H2		Axial force				

( ) is the ratio of acting stress to holding strength.

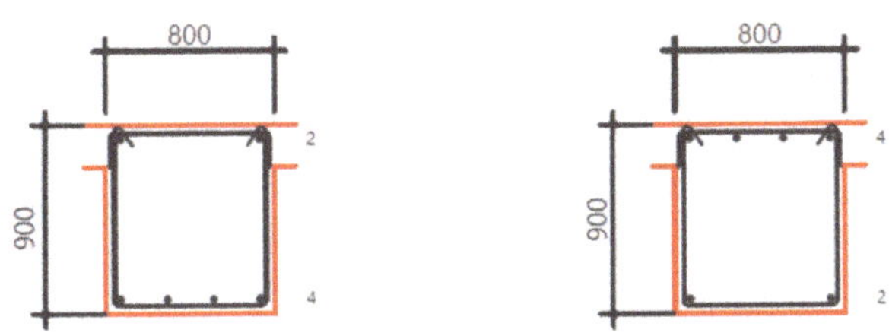

Top: 2-HD25 / Bottom: 4-HD25 / Stirrup: 2-D10@300 — End

Top: 2-HD25 / Bottom: 4-HD25 / Stirrup: 2-D10@300 — Center

Figure 8. Detail of RC girders after the completion of construction.

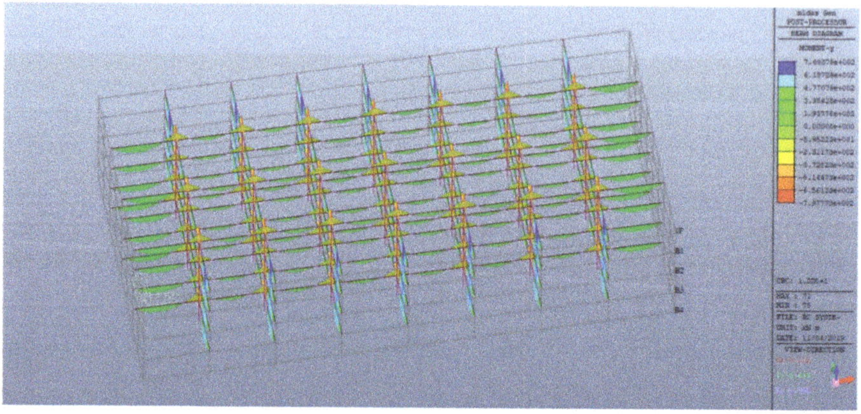

Figure 9. Moment diagram of RC frame structure after the completion of construction.

5.2. *Normal Top-Down Method*

In the illustrative example using the normal top-down method, steel girders and SRC columns were used. The sizes of the used girder and column members of Case #6 are shown in Table 3. As mentioned previously, the first floor can be used for storage space of construction materials. Therefore, the force ratios during construction was higher than those after completion of construction. The moment ratio of girders reached 0.94 and 0.96 at G2 and G4, respectively, during construction. However, they were reduced to 0.42 and 0.44 after construction, respectively. The force ratios of girders of BF increased after completion of construction since the live load increased from 1.5 kN/m^2 to 3.0 kN/m^2. The moment ratios of G2 and G4 were 0.95 and 0.89 after completion of construction, respectively. For columns, the ratio reached up to 0.67 for C2 during construction, but it was reduced to 0.15 after construction. For the results, all cases were structurally stable and the member were designed optimally.

Table 3. Moment size and analyzed force ratios of structural members in the illustrative example using the normal top-down method (Case #6).

Member		Size		Force Type	Ratio of Force to Holding Strength			
					During Construction		After Completion of Construction	
		1F	BF		1F	BF	1F	BF
Girder	G1	H-488 × 300 × 11 × 18	H-600 × 200 × 12 × 20	Moment	655/725 (0.90)	120/320 (0.38)	296/725 (0.41)	210/320 (0.66)
				Shear force	536/1110 (0.47)	97/1373 (0.07)	237/1110 (0.17)	169/1373 (0.12)
	G2	H-588 × 300 × 12 × 20	H-600 × 200 × 12 × 20	Moment	914/978 (0.94)	172/320 (0.54)	411/978 (0.42)	303/320 (0.95)
				Shear force	647/1461 (0.44)	120/1389 (0.0.9)	293/1461 (0.20)	210/1389 (0.15)
	G3	H-700 × 300 × 13 × 24	H-612 × 202 × 13 × 23	Moment	1013/1398 (0.73)	187/515 (0.36)	457/1398 (0.33)	321/515 (0.62)
				Shear force	536/1885 (0.28)	102/1648 (0.06)	245/1885 (0.13)	174/1648 (0.11)
	G4	H-700 × 300 × 13 × 24	H-612 × 202 × 13 × 23	Moment	1348/1398 (0.96)	264/515 (0.51)	608/1398 (0.44)	460/515 (0.89)
				Shear force	710/1885 (0.38)	137/1634 (0.08)	325/1885 (0.17)	235/1634 (0.14)
Column	C1	900 × 900 (H-414 × 405 × 18 × 28)		Axial force	3612/5718 (0.63)		3343/23754 (0.14)	
	C2			Axial force	3807/5718 (0.67)		3522/23754 (0.15)	

( ) is the ratio of acting stress to holding strength.

5.3. Top-Down with MPD Method

For the bottom-up method, the column was installed between MPDs in RC structure while the beam was a steel structure like for the normal top-down method. In the illustrative example of the case 14 using the top-down method with MPDs, the size of the supporting frames used was H-300 × 300 × 10 × 15 mm and micro piles were installed around the existing columns in the basement. The sizes of the girders and columns used were 800 × 900 mm and 900 × 900 mm, respectively. Table 4 shows the force ratio of each member in the example in which the top-down with the MPD method was applied for vertical underground extension. The high moment ratios of 0.96 and 0.89 were found at G2 and G4, respectively. The shear force ratios of girders were not larger than 0.15. The axial fore ratios of MPDs reached 0.75 and 0.81 during construction while those of column to 0.36 and 0.37 after construction. The illustrative example using the top-down method with MPDs was structurally stable, because the force ratios of all structure members were less than 1.0. According to the structural analysis results in the illustrative examples to which the three vertical underground extension methods were applied, all cases were structurally stable.

Table 4. Member size and analyzed force ratios of structural members in the illustrative example using the top-down with the MPD method (Case #14).

Member		Size		Force Type	Ratio of Force to Holding Strength	
		1F & B1F	B2F & B3F		1F & B1F	B2F & B3F
Girder	G1	Existing member size	H-600 × 200 × 12 × 20	Moment	-	210/320 (0.66)
				Shear force		169/1373 (0.12)
	G2	Existing member size	H-600 × 200 × 12 × 20	Moment	-	303/320 (0.95)
				Shear force		210/1389 (0.15)
	G3	Existing member size	H-612 × 202 × 13 × 23	Moment	-	321/515 (0.62)
				Shear force		174/1648 (0.11)
	G4	Existing member size	H-612 × 202 × 13 × 23	Moment	-	460/515 (0.89)
				Shear force		235/1634 (0.14)
Column	C1	900 * 900		Axial force	4148/11645 (0.36)	
	C2			Axial force	4354/11645 (0.37)	
MPD	MPD1	H-300 × 300 × 10 × 15		Axial force	1113/1480 (0.75)	
	MPD2	H-300 × 300 × 10 × 15		Axial force	1204/1480 (0.81)	

( ) is the ratio of acting stress to holding strength.

6. Economic Impacts

The extension costs of each illustrative example were calculated based on the expected amount of used and demolished materials. Table 5 represents the detailed extension cost of the case # 16. The calculated vertical underground extension cost consisted of material and labor costs. Material costs were classified as reinforcement, demolition, earthworks, structural frame construction, and finishing costs. Materials for the reinforcement work included support (EA), additional reinforcement member (m^3), MPD posts (boring: m, H-pile: ton), beams for MPD (H-pile: ton), etc. in the paper. Earthworks were related with retaining walls (CIP), PRD (D800: m), excavation, temporary struts, and so on. Structural frames were mainly composed of slabs, beams, columns, exterior walls, and foundations. To calculate materials cost of each structural member, form (m^2), concrete (m^3), support (m^2), reinforced bar (ton), and deck plate (m^2) were considered.

Table 5. Detailed construction costs of Case #16.

Division of Construction			Unit	Quantity	Unit Price (USD)	Cost (USD)	Sum (USD)
Reinforcement	Support		EA	2	61	164,591	978,365
	Structural reinforcement		M^3	2190	174	380,870	
	MPD Post	Drilling	M	1803	70	125,426	
		H-pile	ton	169	1391	235,130	
	Beam supporting MPDs	H-beam	ton	52	1391	72,348	
Demolition	Raker		Set	0	0	0	85,696
	RC frame		M^3	0	130	0	
	Foundation		M^3	3285	26	85,696	
Earthwork	CIP		Set	1		798,696	1,352,000
	PRD	D800	M	0	261	0	
	Excavation		M^3	18,180	30	553,304	
	Strut		Set	0	0	0	
Structural frame	Slab	Form	M^2	0	45	0	1,001,341
		Concrete	M^3	0	74	0	
		Reinforcing bars	ton	0	957	0	
		Support	M2	0	9	0	
		Concrete in deck	M^3	805	74	59,500	
		Reinforcing bars in deck	ton	25	957	23,913	
		Deck	M^2	4305	32	138,509	
		Edge construction	M	398	130	51,913	
	Beam	Form	M^2	0	48	0	
		Concrete	M^3	0	74	0	
		Reinforcing bars in deck	ton	0	957	0	
		Support	M^2	0	9	0	
		Steel beam	ton	145	1391	201,739	
		Support for steel beam	EA	48	565	27,130	
		Fire insulation	M^2	2939	8	23,001	
	Column	Form	M^2	728	45	32,918	
		concrete	M^3	164	74	12,122	
		H-beam	ton	0	1391	0	
		Reinforcing bars in deck	ton	12	957	11,478	
	Exterior wall	Form	M^3	1550	45	70,087	
		Concrete	M^3	262	74	19,365	
		Reinforcing bars in deck	ton	69	957	66,000	
	Foundation	Concrete	M^3	2299	74	169,926	
		Reinforcing bars in deck	ton	98	957	93,739	
Finish/other			M^2	6571	391	2,571,261	2,571,261
	Labor		Set			898,296	898,296
	Total					6,886,958	6,886,958

Table 6 represents the sustainable vertical underground extension costs of the 24 illustrative examples. Unlike other methods, the top-down with MPD method includes reinforcement costs and does not include demolition costs. Also, the sustainable vertical underground extension cost of an existing building without a basement floor does not include demolition costs. Figure 10 shows the sustainable vertical underground extension cost per extended floor considering the number of basement floors of an existing building (G/L to B3) and the number of extended basement floors (one to four floor extension). In the illustrative examples, to vertically extend the underground space of an existing building without a basement floor, the extension costs using the bottom-up and the normal top-down methods were compared. It is difficult for the bottom-up and the normal top-down methods with PRD and RCD to vertically extend the underground space of an existing building with more than three basement floors. Therefore, they were not included when analyzing and comparing the economic impacts of vertical underground extension methods. In all examples using the three methods, the larger the number of basement floors of existing buildings, the greater the cost per extended basement floor. The larger the number of basement floors of existing buildings, the higher the cost of demolition and reinforcement works. In particular, the cost of demolition work was strongly related to the number of basement floors of existing buildings. Also, the larger the number of vertically extended basement floors, the dramatically less the extension cost per extended floor in all examples.

Table 6. Extension costs of underground vertical extension methods in the 24 illustrative examples.

Case No.	Extension Cost (USD 1K)						
	Material Cost					Labor Cost	Total Cost
	Reinforcement	Demolition	Earthwork	Structural Frame	Finish/Other		
1	–	–	997	717	857	386	2957
2	–	–	870	757	857	372	2856
3	–	–	1473	1179	1714	655	5021
4	–	–	1282	1117	1714	617	4730
5	–	–	1950	1642	2571	924	7087
6	–	–	1696	1477	2571	862	6606
7	–	–	2464	2104	3428	1199	9196
8	–	–	2152	1840	3428	1113	8533
9	–	1082	1473	1179	1714	817	6264
10	–	1082	1282	1117	1714	779	5974
11	887	86	962	670	1714	648	4967
12	–	1082	1950	1642	2571	1087	8331
13	–	1082	1696	1477	2571	1024	7850
14	978	86	1352	1001	2571	898	6887
15	–	1082	2464	2104	3428	1362	10,440
16	–	1082	2152	1840	3428	1275	9777
17	1071	86	1784	1332	3428	1155	8857
18	–	2092	1950	1642	2571	1238	9492
19	–	2092	1696	1477	2571	1175	9011
20	1524	86	1125	670	2571	896	6873
21	–	2092	2464	2104	3428	1513	11,601
22	–	2092	2152	1840	3428	1427	10,939
23	1616	86	1558	1001	3428	1153	8842
24	2162	86	1331	670	3428	1152	8828

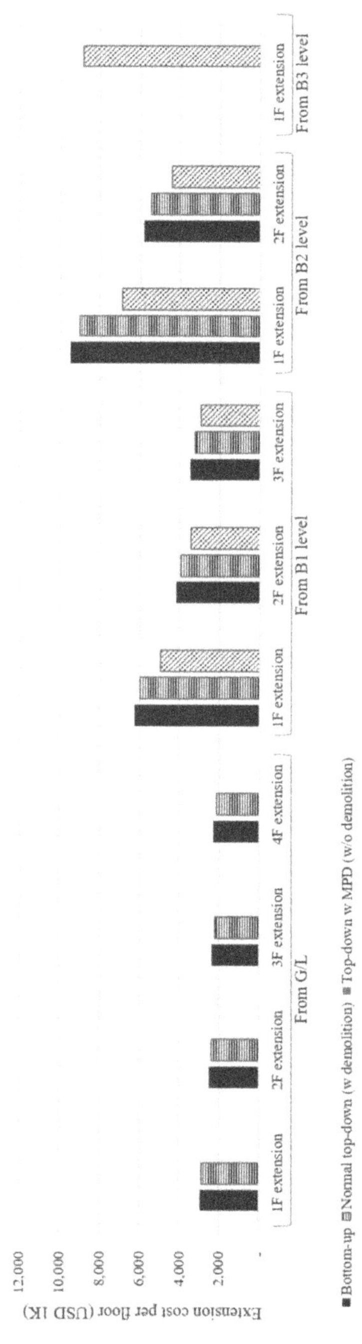

Figure 10. Extension cost per floor considering the number of existing basement floors (G/L to B3) and vertically extended basement floors (B1F to B4F extension).

As the results, the top-down with MPD method showed the lowest extension cost per extended basement floor under the same conditions, followed by the normal top-down and the bottom-up methods. In particular, the vertical extension cost using the top-down with MPD method, for two basement floor extensions of the existing building with two basement floors, was approximately 76% of that using the bottom-up method. As the number of vertically extended basement floors increased, the difference between extension costs of the underground extension methods decreased. Unlike the bottom-up method, the normal top-down method had relatively low extension costs for earthworks and structural framing for temporary columns, support frames, and so on. The demolition cost for the vertical underground extension using the top-down with MPD method was lower than those of other two methods, but the cost of reinforcement work was additionally considered unlike other methods due to the additional MPD installation. Also, the cost of reinforcement work was less than that of demolition work. For vertical underground extension of an existing building with the same number of basement floors, the extension cost per extended basement floor decreased as the number of extended basement floors increased. For example, the costs per basement floor for vertically extending two and three basement floors of an existing building with one basement floor were 68% and 57% of that for extending one basement floor, respectively.

Earthwork cost of the top-down with MPD method was the lowest, followed by the normal top-down and bottom-up methods. The difference in earthwork costs for the bottom-up and top-down methods was related to the cost for installing temporary struts. The normal top-down method additionally considered PRD to calculate the earthwork cost, unlike the top-down with MPD method, while the excavation cost of the top-down with MPD method was lower than those of other methods. Cost of structural frames of the top-down with MPD method was also the lowest, followed by the normal top-down and bottom-up methods. Although cost for foundations of each method was the same, cost for columns and exterior walls of the top-down with MPD method was lower than those of other methods. The amount of newly constructed columns and exterior walls of the top-down with MPD method was smaller than those of other two methods due to absence of demolition works. In particular, for the normal top-down methods, H-piles were used for reinforcing existing columns.

7. Discussion and Conclusions

This paper analyzed the economic impacts of three vertical extension methods for basement floors of existing buildings, which were the bottom-up, normal top-down, and top-down with MPDs methods. Extension cost of the three methods to sustainably expand underground spaces of existing buildings were calculated based on the quantity of used or demolished materials, their unit cost, and the assumed labor cost rate. A process to select an appropriate vertical underground extension method among them was also proposed by considering construction site conditions, such as the number of existing basement floors of buildings and extended basement floors, and whether demolition work is required. The top-down method with MPDs can vertically extend the underground space using micro piles and temporary support frames without demolishing the existing basement floors of buildings, unlike other extension methods that need to demolish the existing underground structure. Micro piles and temporary support frames were used as reinforcement elements for existing and extended underground space and retaining walls.

Based on the three construction methods, the number of basement floors in existing buildings, and the number of extended basement floors, the economic impacts were analyzed by calculating and comparing the underground vertical expansion costs of 24 illustrative examples. When an underground space of the existing one or two basement floors exists, from the cost-oriented perspective, the existing underground structure should be appropriately reinforced by using micro piles, and then two or more layers should be expanded by using the underground expansion method. This might be beneficially coordinated according to the characteristics of projects. In addition, when an underground space of the existing three basement floors exists, the construction cost is estimated to be significant, because the area to be enlarged when the underground vertical extension is constructed at the underground fourth floor is relatively small. In other words, the higher the number of underground floors of existing buildings and the greater the number of underground vertical expansions, the more economically advantageous the top-down with MPDs method is.

Financial analysis does not guarantee the structural stabilities of buildings with underground spaces extended by using these construction methods. Therefore, practitioners should simultaneously consider various issues, such as existing soil and bearing capacity of existing buildings, relevant building regulation or codes, and policies. For example, because of soil conditions on sites, it can be difficult to apply top-down methods or install MPD for support existing underground structures, which are relatively more slender than other types of piles.

The proposed extension processes through considering their structural stabilities and economic impacts help practitioners to select appropriate construction techniques and reduce costs and risks. They will provide opportunities to sustainably and efficiently expand underground space of existing buildings in congested urban areas without dismantling whole buildings. Consequently, it is expected that the amount of the construction and demolition (C&D) waste generated during the demolition phase of existing buildings and the construction phase of new buildings will be reduced. However, the proposed process was not yet applied to actual projects. Therefore, in the future, the strengths and weaknesses of the proposed process will be analyzed based on the implementation of sustainable underground vertical extension methods on actual sites.

Author Contributions: S.-y.S. made substantial contributions to conceptualization of the study, acquisition, analysis, and drafting and revising the article; B.L. helped to collect relevant data and analyze the result; J.W. proposed the research methodology and reviewed and edited the manuscript. All authors have read and agreed to the published version of the manuscript.

Acknowledgments: This work is supported by the Korea Agency for Infrastructure Technology Advancement (KAIA) grant funded by the Ministry of Land, Infrastructure and Transport (Grant No. 18EREP B099826-04).

Conflicts of Interest: The authors declare no conflict of interest.

References

1. Jung, S.J.; Kim, S.-K.; Seo, S.Y. Structural safety of the building constructed by top-down method corresponding to earth pressure distribution and floor system. *J. Archit. Inst. Korea Plan Des.* **2017**, *33*, 3–10.
2. Volchko, Y.; Norrman, J.; Ericsson, L.O.; Nilsson, K.L.; Markstedt, A.; Öberg, M.; Mossmark, F.; Bobylev, N.; Tengborg, P. Subsurface planning: Towards a common understanding of the subsurface as a multifunctional resource. *Land Use Policy* **2020**, *90*, 104316. [CrossRef]
3. Volchko, Y.; Norrman, J.; Ericsson, L.O.; Nilsson, K.L.; Markstedt, A.; Öberg, M.; Mossmark, F.; Bobylev, N. Underground space as an urban indicator: Measuring use of sub- surface. *Tunn. Undergr. Sp. Tech.* **2016**, *55*, 40–51.
4. Bobylev, N. Comparative analysis of environmental impacts of selected underground construction technologies using the analytic network process. *Autom. Constr.* **2011**, *20*, 1030–1040. [CrossRef]
5. Chen, Z.; Su, L.; Zhang, C. Research on the Synergy Degree of Aboveground and Underground Space along Urban Rail Transit from the Perspective of Urban Sustainable Development. *Sustainability* **2016**, *8*, 934. [CrossRef]
6. Bobylev, N. Mainstreaming sustainable development into a city's Master plan: A case of Urban Underground Space use. *Land Use Policy* **2009**, *26*, 1128–1137. [CrossRef]
7. Yang, X.; Chen, Z.; Cai, H.; Ma, L. A Framework for Assessment of the Influence of China's Urban Underground Space Developments on the Urban Microclimate. *Sustainability* **2014**, *6*, 8536–8566. [CrossRef]
8. Ro, Y.-C.; Lee, C.-S. Selection of retaining wall system for underground parking lots expansion of apartments. *J. Korea Inst. Constr. Eng. Manag.* **2008**, *9*, 99–107.
9. Kim, K.M.; Rhim, H.C.; Lee, K.J. Development of Underground Space Underneath Existing Buildings. In Proceedings of the 2010 Conference of Architectural Institute of Korea, Seoul, Korea, 20–23 October 2010; Architectural Institute of Korea: Cheongju, Korea, 2010; pp. 153–154.
10. Bing, C. Floating underground extension method—Remodeling case 2. *Remodeling* **2012**, *46*, 13–20.

11. Jang, D.; Park, D.; Kim, Y.; Lim, H. Research on the Development of FUSEM Method for Seoul City Main Hall. In Proceedings of the Conference of Architectural Institute of Korea, Gwangju, Korea, 22–25 October 2012; Architectural Institute of Korea: Gwangju, Korea, 2010; pp. 513–514.
12. Lee, J.; Bing, C. Remodeling case—Chungdam Chunggu Apartment. *Build. Construct.* **2012**, 42–51.
13. Kim, T.H.; Bang, J.S.; Lim, C.W. A Study on the Basement Extension Construction Method Using Existing Piles. In Proceedings of the 2014 Conference of Architectural Institute of Korea, Seoul, Korea, 25 April 2014; Architectural Institute of Korea: Seoul, Korea, 2014; pp. 407–408.
14. Park, D.-S.; Lew, Y.-K.; Choi, K.; Lee, J. Introduction of floating underground space extension method (FUSEM) for preservation and continuous utilization of old Seoul city hall. *J. Korea Concr. Inst.* **2013**, *25*, 44–48.
15. Kim, S.K.; Lee, J.E.; Kim, T.W.; Koo, J.M.; Jung, S.; Seo, S.Y. Construction Technology and Structural Safety Assessment for Expansion of Underground Parking Lot of Apartment Complex. In Proceedings of the 2016 Conference of Architectural Institute of Korea, Architecutral Institute of Korea, Busan, Korea, 4–6 October 2016; pp. 1471–1472.
16. Jung, S.J.; Kim, J.H.; Lee, B.H.; Hwang, K.S.; Seo, S.Y. Establishment of a Technology for Earthwork Safety at New Construction/Extension of Underground Parking Lot in Building. In Proceedings of the 2018 Conference of Korea Concrete Institute, Korea Concrete Institute, Changwon, Korea, 1 May 2018; pp. 797–798.
17. Attar, A.; Boudjakdji, M.A.; Bhuiyan, N.; Grine, K.; Kenai, S.; Aoubed, A. Integrating numerical tools in underground construction process. *Eng. Constr. Arch. Manag.* **2009**, *16*, 376–391. [CrossRef]
18. Ha, T.; Lee, S. Advanced Construction Stage Analysis of High-Rise Building Considering Creep and Shrinkage of Concrete. In Proceedings of the 2013 World Congress on Advances in Structural Engineering and Mechanics (ASEM13), Jeju, Korea, 8–12 September 2013; International Association of Structural Engineering and Mechanics: Jeju, Korea, 2013; pp. 2139–2147.
19. Seo, S.; Lee, R.; Construction, S.H. Permanent Earth Retaining Composite Frames for Constructing Underground Structure. In Proceedings of the Conference of Architecutal Institute of Korea, Seoul, Korea, 24–25 October 2005; Architectural Institute of Korea: Jecheon, Korea, 2005; pp. 41–68.
20. Thompson, J.; Zadoorian, C. A Case Study for Top-Down and Construction Methodology for a High-Rise Development in Los Angeles, California. In Proceedings of the SEAOC Convention, Big Island, HI, USA, 23–27 September 2008; SEAOC: Sacramento, CA, USA, 2008; pp. 1–8.
21. Archidata Design. Eart Retaining Wall. Available online: http://www.archidata.co.kr/NewWin/NewWin.asp?VT=Design&selID=54599&ddcodeid=2291 (accessed on 9 January 2020).
22. Ha, S.; Choi, C.; Park, M.; Hwang, K.; Lee, L. Development of New Construction Method of ESD and BRD. In Proceedings of the Conference of Archietctural Institute of Korea, Jecheon, Korea, 24–25 October 2005; Architectural Institute of Korea: Jecheon, Korea, 2005; pp. 305–308.

MDPI
St. Alban-Anlage 66
4052 Basel
Switzerland
Tel. +41 61 683 77 34
Fax +41 61 302 89 18
www.mdpi.com

Sustainability Editorial Office
E-mail: sustainability@mdpi.com
www.mdpi.com/journal/sustainability

www.ingramcontent.com/pod-product-compliance
Lightning Source LLC
Chambersburg PA
CBHW062105220926
51850CB00010B/2895
9783036504841